T0295762

THE POWER-SAVING BEHAVIOR OF HOUSEHOLDS

HOW SHOULD WE ENCOURAGE POWER SAVING?

GREEN RESEARCH, DEVELOPMENTS, AND PROGRAMS

Additional books in this series can be found on Nova's website under the Series tab.

Additional e-books in this series can be found on Nova's website under the eBooks tab.

GREEN RESEARCH, DEVELOPMENTS, AND PROGRAMS

THE POWER-SAVING BEHAVIOR OF HOUSEHOLDS

HOW SHOULD WE ENCOURAGE POWER SAVING?

KENICHI MIZOBUCHI

AND

HISASHI TANIZAKI

nova
science publishers
New York

Copyright © 2018 by Nova Science Publishers, Inc.

Additional color graphics may be available in the e-book version of this book.

Library of Congress Cataloging-in-Publication Data

ISBN: 978-1-53613-173-4

Published by Nova Science Publishers, Inc. † *New York*

CONTENTS

PREFACE

This book is about the empirical analysis of household electricity saving behavior. In particular, we focus on effective methods to promote energy saving behavior and the effectiveness of energy-saving equipment. After the Great East Japan Earthquake of 2011, 52 of Japan's nuclear power plants temporarily stopped. Prior to the Fukushima accident, about 25% of Japan's total electricity supply amount depended on nuclear power. Therefore, the resulting power shortage has become a serious problem, especially in summer and winter. In this document, the authors focus on several policy instruments that encourage energy saving behavior such as economic incentive (increase in electricity price and compensation), public electricity saving request, comparative feedback, social norms, and verify their effect. Furthermore, the authors will conduct a quantitative economic analysis based on these data sets using randomly obtained data as well as summary data announced after 2011 Great East Japan Earthquake. The authors then examine how households respond to policy measures to save electricity. The result of this book is to clarify to what extent the power saving policy targeted at homes was effective, and it is useful for considering what kind of policy measures (including a mixed policy) should be adopted according to future goals. The authors especially recommend this book to researchers and environmental energy policy-makers, but also target readers interested in Japan's energy saving issues.

AIM OF THE BOOK

In this way, there are two objectives (reduction of greenhouse gas emissions and response to power shortage) to encourage Japanese households to save energy. However, in response to these objectives, the policies that the government has taken so far have mainly been dependent on the voluntary action of the household sector. Regarding the reduction of greenhouse gas emissions, the Japanese government has set a reduction target of 26% by 2030 compared to 2013, among which the reduction target of the household sector is set at 39.3%, which is very high. Several nuclear power plants are still undecided regarding the time of restart, and nuclear power plants that determine decommissioning furnaces are also emerging. Under such circumstances, the Japanese household sector will increasingly be required to save electricity in the future.

The purpose of this book is to discuss the effective methods for encouraging energy conservation behavior in the household sector, using field experiments and econometric methods. The organization of this book is as follows. In Chapter 2, we examine the factors affecting the electricity demand of the household sector in Japan, mainly focusing on changes in prices and income. Specifically, we estimate the demand system model of the household sector and analyze electricity demand from the estimates of price and expenditure elasticities. We also examine whether structural changes occur in the energy demand due to the Great East Japan Earthquake. In Chapter 3, we analyze the effect of the government's request to save electricity after the Great East Japan Earthquake using econometric methods. We also examine areas where power-saving requests have not been issued, and whether or not the electricity conservation behavior of households exists. Chapters 4 to 6 are researches on promoting energy-saving behavior among households, using field experiments. In Chapter 4, we examine whether incentives promote energy-saving behavior economically. Specifically, by setting stepwise remuneration according to the power-saving rate, we analyze whether a change in power-saving behavior has occurred. Chapter 5 examines whether psychological factors promote energy-saving behavior in addition to economic incentives. Specifically, we divided the participating households into several groups: i) compensation based on the power-saving rate; ii) comparative feedback (comparison with the power-saving situation of other households in the group); and iii) the control group. We also examine whether a significant difference is confirmed in the average power-saving rate of each group, and verify the magnitude of the marginal cost of energy saving. In Chapter 6, we examine whether there is a power-saving effect by using energy-saving home appliances. Specifically, we compare the average energy-saving rate of households that have replaced high-energy consuming air conditioners with energy-saving ones to those that have not replaced them within the past two years. In doing so, we use a combination of difference-in-differences and propensity score matching method for a more accurate comparison. Chapter 7 provides a summary of this book and directions for future energy-saving policies.

Chapter 1

WHY DO JAPANESE HOUSEHOLDS NEED POWER-SAVING BEHAVIOR?

ABSTRACT

This chapter discusses why Japanese households need energy-saving behavior, from the demand- and supply-side perspectives. Two major reasons have been identified for saving energy: i) reduction of greenhouse gas emissions; and ii) response to insufficient power supply. In the Japanese household sector, policies have been implemented to promote energy saving through voluntary means, but not through regulatory methods. However, it cannot be said that these policies had no significant effect. In this chapter, we discuss the necessity of energy saving by analyzing the problems related to the changes in electricity demand in the household sector and electricity supply, using actual data.

Keywords: energy saving, households, CO_2 emission

1. INTRODUCTION

1.1. Demand/Residential Demand for Electricity

In 2015, the Japanese household sector accounted for approximately 14.6% of the country's carbon dioxide emissions, although this share was less than those of the industrial (33.5%) and transportation (17.4%) sectors. However, it is noteworthy that from 1990 to 2015, household emissions surged by 37.4%, and continue to do so every year, in sharp contrast to industrial (-18.1%) as well as transportation (3.5%) sector emissions (see Figure 1.1.).[1] The primary reason for this change is said to be the increasing demand for household electricity. Figure 1.2 shows the trends in energy consumption by fuel type. The Figure illustrates that electricity accounted for approximately 49.6% of energy consumption in 2015, followed by oil (19.6%) and city gas (19.8%); more specifically, the demand for electricity has increased significantly since 1990.

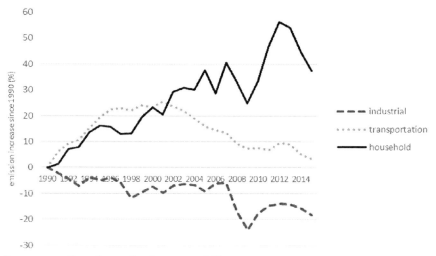

Data source: Greenhouse Gas Inventory Office.

Figure 1.1. CO$_2$ emission trends (1990–2015).

---

[1] Greenhouse Gas Inventory Office, Japan: http://www-gio.nies.go.jp/.

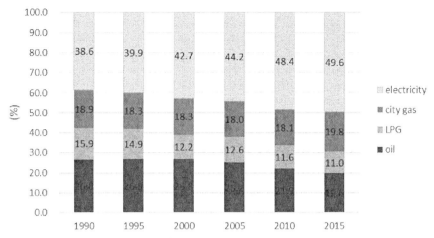

Data source: The Energy Data and Modeling Center (EDMC).

Figure 1.2. Trends in the ratio of energy consumption by fuel type.

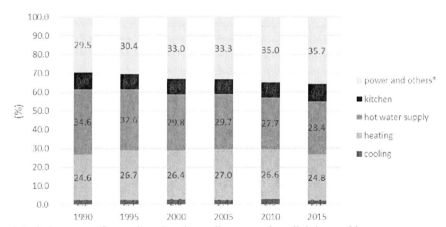

* Includes power from other electric appliances such as lighting, refrigerator, vacuum cleaner, and television.

Data source: The Energy Data and Modeling Center (EDMC).

Figure 1.3. Trends in the ratio of energy consumption by application.

Figure 1.3 shows the trends in energy consumption by a household application. The Figure demonstrates that from 1990 to 2015, the proportion of energy to "power and others" has been increasing. "Power and others" mainly consists of electric power necessary for the use of home appliances (televisions, refrigerators, and so on). In other words, it is clear that the penetration rate of household appliances has increased during the past 25 years. Figure 1.4 and Figure 1.5 show the trends in the penetration rate of household electrical appliances. According to Figure 1.4, the penetration rate of major household appliances such as air conditioners, televisions, and refrigerators is nearly 100%. Meanwhile, as of 1990, the penetration rate of personal computers and clothes dryers, which had hardly spread, was quite insignificant. However, it has been rapidly increasing since then. These household electric appliances consume significant electricity, and thus, their contribution to household electricity consumption increase is large.

Furthermore, the rise in ownership rate of household appliances may have an impact on household electricity consumption. According to Figure 1.5, the ownership rates of air conditioner, personal computer, and warm water washing toilet seat have increased about 1.2, 9.1, and 6.2 times, respectively, during the past 25 years. In addition, the diversification and wide usage of home electric appliances such as dishwashers and automatic vacuum cleaners as well as electronic devices such as smartphones and tablet terminals have rapidly increased the electricity consumption of households in recent years.

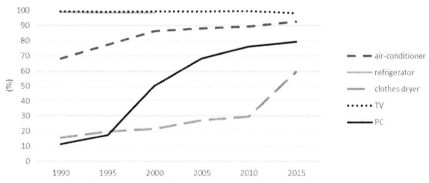

Data source: The Energy Data and Modeling Center (EDMC).

Figure 1.4. Trends in the penetration rate of home electric appliances.

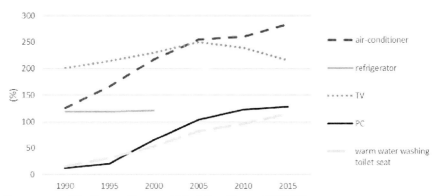

Data source: The Energy Data and Modeling Center (EDMC).

Figure 1.5. Trends in the ownership rate of home electric appliances.

Thus, the demand for electricity can be said to have increased due to the popularization and diversity of household electric appliances; however, there also factors that work to reduce electricity consumption, such as energy-saving options for home appliances. Figure 1.6 shows the transition of power consumption in general air conditioners. The figure demonstrates that the power consumption of air conditioners has been consistently decreasing, and in 2015, the electricity consumption was nearly half of that in 1995. In Japan, the government's Top Runner Program[2] promptly encouraged the development and dissemination of energy-saving products. Therefore, it is considered that the rise in electricity consumption is suppressed by energy saving of home appliances.

---

[2] The purpose of this program is to promote energy-efficient electric appliances and automotive vehicles. Under this program, energy-efficiency targets are set based on the value of the most energy-efficient products on the market at the time of the value-setting process, and all machinery and equipment covered by the program should exceed these targets. If a company continues to sell products that fall below the targets, the company's name will be published and penalty charges will be imposed.
(http://www.enecho.meti.go.jp/policy/saveenergy/toprunner2011.03en-1103.pdf).

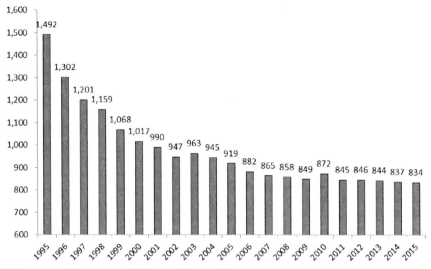

Data source: Resources and Energy Agency, Japan.

Figure 1.6. Annual power consumption of air conditioner from 1995–2013 (kWh).

1.2. Supply Side of Electricity

After the incident of the Fukushima Daiichi Nuclear Power Plant, which took place due to the occurrence of the Great East Japan Earthquake of March 2011, Japan's electricity supply underwent a major transition. Figure 1.7 shows the transition of electricity supply in Japan. After the incident, Japan's nuclear power plants, which accounted for about 25% of Japan's electricity supply, discontinued operations because of concerns about safety, and thus, the supply of electricity was significantly reduced. This discontinuance increased the proportion of thermal power generation (67% in 2010 and 90% in 2015), and consequently, the amount of fossil fuel used and CO_2 emissions increased. On the other hand, due to the reduction of electricity supply itself, problems of power shortage occurred, especially when the demand for electricity increased in the summer and winter.

Therefore, in order to maintain the balance between the demand and supply, the Japanese government restricted and requested that all departments reduce power consumption. However, unlike the industrial sector, it was difficult to compel the household sector to save electricity. Thus, the Japanese government urged the household sector to save electricity by providing concrete methods for saving electricity and presenting power-saving targets. There are three studies that examined the effect of the government's request to conserve electricity for the household sector after the earthquake. Nishio and Ofuji (2013) surveyed 1,500 households through a web-based questionnaire. The results revealed that in the summer of 2012, compared to the summer of 2010, before the earthquake, about 10% electricity was saved. Tanaka and Ida (2013) conducted a web-based questionnaire survey in the Kanto and Kansai regions to examine the voluntary energy-saving behavior of the household sector after the earthquake. Their findings confirmed a high level of voluntary energy-saving behavior; however, they concluded that it is difficult to maintain this behavior for a long period. Arikawa et al. (2014) examined the relationship between individual electricity demand and nuclear acceptance after the Fukushima Dai-ichi nuclear disaster; using web-based survey data, which included about 800 households, they demonstrated that opponents of nuclear power used electrical appliances less intensively at home and reduced their electricity consumption during the power shortage period. On the other hand, supporters of nuclear power had high energy demands and did not save on power consumption.

The request for energy saving is a policy that the Japanese government has principally carried out, and in order to stabilize the electricity supply, it is necessary to encourage companies and households to maintain a certain level of voluntary energy-saving behavior (as some obligations for some companies). In Japan, there are 10 electric power companies that supply electricity monopolistically by region (i.e., Hokkaido Electric Power Company, Tohoku Electric Company, Tokyo Electric Company, Hokuriku Electric Company,

Tokai Electric Company, Kansai Electric Company, Tyugoku Electric Company, Shikoku Electric Company, Kyusyu Electric Company, and Okinawa Electric Company). Figure 1.8 shows the areas where each electric power company supplies electricity. The level of the government's request to save electricity depends on the size of the nuclear power plant owned by the electric power company in each region. The Okinawa electric power company does not hold a nuclear power plant; therefore, it did not become a target for the power-saving request. Table 1.1 shows the level of power-saving requests issued to each region in the summer of 2011 and 2012. The energy-saving rate is based on the amount used in 2010. The power-saving request was implemented from 1 July to 30 September, from 9:00 to 20:00 on weekdays. The Table shows that in 2011, Tokyo Electric Company's power supply area was requested to conserve 15% compared to 2010. Even in October 2017, only five nuclear power plants have been operating, and there is a possibility that power shortage may recur due to sudden changes in the temperature in summer and winter.

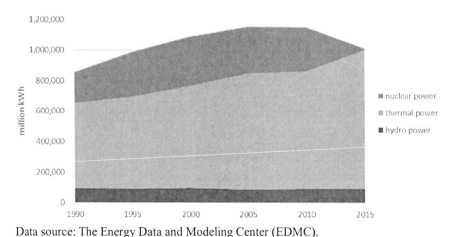

Data source: The Energy Data and Modeling Center (EDMC).

Figure 1.7. Trends of generated electric energy.

Figure 1.8. Service areas of 10 electric power companies in Japan.

Table 1.1. Power-saving request level by region in summer

	2011	2012
Hokkaido	-	7%
Tohoku	15%	-
Tokyo	15%	-
Hokuriku	-	-
Tokai	-	-
Kinki	10%	10%
Tyugoku	-	-
Shikoku	-	5%
Kyusyu	-	10%
Okinawa	-	-

Source: http://www.kantei.go.jp/jp/headline/summer2012_denryoku.html#c2_2.

Chapter 2

Estimation of the Demand System Model of the Japanese Household Sector: Have There Been Structural Changes in Energy Demand Elasticities during the Great East Japan Earthquake?

Abstract

After the Great East Japan Earthquake in March 2011, the household sector's energy demand, which had witnessed an upward trend, began to decrease. In this chapter, we estimated the household demand system model by using the Japanese household expenditure data from January 1975 to August 2017. Thereafter, from the estimated parameters of the demand system model, we estimated the price and expenditure elasticities of demand for each item. As a result of the estimation, we found that structural changes occurred in household demand due to the disaster. The value of expenditure elasticity became positive and significant (0.55) after the disaster. On the other hand, households' response to energy prices has increased than ever before, as price elasticity became negative and significant and was even larger. This result may indicate the effectiveness of economic incentives in the future for energy shortage and global warming countermeasures (this chapter is a substantial revision of Tanizaki and Mizobuchi, 2014).

Keywords: AIDS model, price elasticity, expenditure elasticity, structural change

2.1. INTRODUCTION

As shown in Chapter 1, energy expenditure by Japanese households has continued to increase. In particular, the proportion of expenditure to electric power is increasing, as the diversification of Internet-related products such as personal computers and tablets has progressed, in addition to the rapid spread of home appliances and increased ownership rates of household appliances. This may be attributed to the Great East Japan Earthquake that occurred in March 2011. After the earthquake, nuclear power plants nationwide discontinued operations, causing power shortage problems in various regions. Consequently, activities to promote power saving spread nationwide.

In order to analyze energy conservation behavior of households, it is necessary to first examine household consumption behavior. Accordingly, the income and price elasticities of demand for each good are used for the analysis. However, previous studies estimated the elasticity of energy demand using time series data for a single item (Okajima and Okajima 2013). Meanwhile, since it is believed that the change in energy demand also affects the demand for other goods, it is necessary to consider the influence on other goods, using the demand system model.

A commonly used demand system model is the Almost Ideal Demand System (hereafter, AIDS) model proposed by Deaton and Muellbauer (1980), which provides the second-order approximation to an arbitrary demand system and satisfies perfect aggregation conditions over consumers. In addition, using the AIDS model we can easily test the constraints (i.e., homogeneity, symmetry, and negativity) in the classical demand theory. The AIDS model has been extensively applied in various empirical studies on demand analysis and fields such as agricultural economics, demographic economics, development economics, environmental economics, health economics, industrial organization, monetary economics, and urban economics (e.g., Rossi 1988; Tiffin and Aguiar 1995; Filippini 1995; Oladosu 2003; Hashimoto 2004; Mizobuchi 2008).

However, many previous empirical studies using the AIDS model analyzed point estimates of elasticity calculated from estimated parameters, and the significance of the elasticity itself was not a problem. If the elasticity is not significant, the analysis using that value also has no meaning. Tanizaki and Mizobuchi (2014) proposed an estimation method using the bootstrap method to obtain statistical information on the elasticity of the AIDS model (i.e., the standard error, p-value, confidence interval, and so on). In this chapter, using the method of Tanizaki and Mizobuchi (2014), we estimate the AIDS model from household expenditure data and statistically analyze the magnitude of price elasticity and expenditure elasticity of energy demand. Furthermore, we clarify whether structural changes in household demand behavior occurred due to the Great East Japan Earthquake, and if so, then what kind of structural changes occurred.

This chapter is structured as follows. Section 2.2 provides the model specification, in which AIDS and LA-AIDS models are discussed. Section 2.3 presents the estimation method of the AIDS model proposed by Tanizaki and Mizobuchi (2014). Section 2.4 examines an empirical analysis using Japanese household expenditure data from January 1975 to August 2017, and Section 2.5 estimates the price and expenditure elasticities of energy demand. We also test the structural changes due to the Great East Japan Earthquake. Section 2.6 concludes the chapter.

2.2. THE DEMAND SYSTEM MODEL

2.2.1. AIDS and LA-AIDS Models

The AIDS model (see Deaton and Muellbauer 1980) is derived from a PIGLOG (i.e., Price Independence Generalized Logarithmic) type of the indirect cost function. This cost function has several parameters and can be interpreted as the second-order approximation to any cost function. Shephard's lemma gives us a set of demand equations which can be considered as an approximation of any demand system. Let $w_{i,t}$ be the share of the ith good, i.e., $w_{i,t} = X_{i,t}/X_t$, where $X_{i,t}$ indicates the expenditure of the ith good at time t and X_t represents $X_t = \sum_{i=1}^{M} X_{i,t}$. Let M be the number of $i = 1$ goods, $P_{i,t}$ be the price of the ith good and P_t be the price index. The AIDS model is expressed as follows:

$$w_{i,t} = \alpha_i + \beta_i ln \frac{X_t}{P_t} + \sum_{j=1}^{M} \gamma_{ij} ln p_{j,t} + u_{i,t}, \qquad (2.1)$$

for $i = 1,2, ..., M$ and $t = 1,2, ..., T$, where $u_{i,t}$ indicates the error term in the ith equation, and the variance–covariance structure is usually assumed to be as follows:

$$E[u_{i,t}u_{j,s}] = \begin{cases} \sigma_{ij}, & \text{if } t = s, \\ 0, & \text{otherwise}. \end{cases}$$

In the case where the error term is serially correlated, the assumption of $E[u_{i,t}u_{j,s}] = 0$ for $t \neq s$ is violated. In Section 2.4, we discuss the bootstrap method in the case where the error term is serially correlated, which is called the Moving Block Bootstrap (MBB) method, and the bootstrap method in the case where the domain of the dependent variables is not from $-\infty$ to ∞, which is called the Pairwise Bootstrap (PB) method.

Equation (2.1) is called the AIDS model when lnP_t is defined as follows:

$$lnP_t = \alpha_0 + \sum_{k=1}^{M} \alpha_k lnp_{k,t} + \frac{1}{2}\sum_{k=1}^{M} \sum_{j=1}^{M} \gamma_{kj} lnp_{k,t} ln\, p_{j,t}.$$

(2.2)

Instead of Equation (2.2), however, most of the empirical studies utilize the following:

$$lnP_t = \sum_{k=1}^{M} w_{k,t} lnp_{k,t},$$ (2.3)

where P_t is called the Stone price index and it is known as an approximation of P_t in (2.2). Hereafter, in this paper, (2.2) is called the original price index to distinguish from the Stone price index. The AIDS model, which utilizes (2.3), is known as the LA-AIDS model.

Deaton and Muellbauer (1980) mention that the Stone price index (2.3) is close to the original price index (2.2). This LA-AIDS model is widely used because of simplicity of the estimation procedure. However, it is well known that the LA-AIDS model yields biased parameter estimates.[3]

From the demand theory, there are some constraints in the AIDS model (2.1), which are as follow;

1) Additivity $\sum_{i=1}^{M} \alpha_i = 1, \sum_{i=1}^{M} \beta_i = 0,$ and $\sum_{i=1}^{M} \gamma_{ij} = 0,$

---

[3] Pashardes (1993) shows that the LA-AIDS model results in an omitted variable problem, which is a source of the biased parameter estimates. This problem is discussed in Buse (1994, 1998). Through some Monte Carlo experiments, Buse (1994) finds that not only SUR but also 3SLS (three stage least squares) yield inconsistent parameter estimates because of the omitted variable problem.

2) Homogeneity $\sum_{j=1}^{M} \gamma_{ij} = 0$,

3) Symmetry $\gamma_{ij} = \gamma_{ji}$.

4) Negativity   The Slutsky matrix (or equivalently, the sub-situation matrix) is negative semidefinite.

For both AIDS and LA-AIDS models, practically, we estimate the first $M-1$ equations. Substituting two additivity constraints $\alpha_M = 1 - \sum_{k=1}^{M-1} \alpha_k$ and $\gamma_{Mj} = -\sum_{k=1}^{M-1} \gamma_{kj}$ into (2.2), we can rewrite the price index as follows:

$$\ln P_t = \alpha_0 + \sum_{k=1}^{M-1} \alpha_k \ln p_{k,t} + \alpha_M \ln p_{M,t}$$
$$+ \frac{1}{2}\sum_{k=1}^{M-1}\sum_{j=1}^{M} \gamma_{kj} \ln p_{k,t} \ln p_{j,t} + \frac{1}{2}\sum_{j=1}^{M} \gamma_{Mj} \ln p_{M,t} \ln p_{j,t}$$
$$= \alpha_0 + \sum_{k=1}^{M-1} \alpha_k \ln p_{k,t} + (1 - \sum_{k=1}^{M-1} \alpha_k)\ln p_{M,t}$$
$$+ \frac{1}{2}\sum_{k=1}^{M-1}\sum_{j=1}^{M} \gamma_{kj} \ln p_{k,t} \ln p_{j,t}$$
$$+ \frac{1}{2}\sum_{j=1}^{M}(-\sum_{k=1}^{M-1}\gamma_{kj}) \ln p_{M,t} \ln p_{j,t}. \tag{2.4}$$

Moreover, substituting (2.4) into (2.1), the AIDS model is estimated as follows:

$$w_{i,t} = (\alpha_i - \beta_i \alpha_0) + \beta_i \ln X_t + \sum_{j=1}^{M-1}(\gamma_{ij} - \beta_i \alpha_j)\ln \frac{p_{j,t}}{p_{M,t}}$$
$$+ (\sum_{j=1}^{M} \gamma_{ij} - \beta_i)\ln p_{M,t} - \frac{1}{2}\beta_i \sum_{k=1}^{M-1}\sum_{j=1}^{M} \gamma_{kj} \ln \frac{p_{k,t}}{p_{M,t}} \ln p_{j,t}$$
$$+ u_{i,t} \tag{2.5}$$

for $i = 1, 2, ..., M-1$, where the two additivity constraints $\alpha_M = 1 - \sum_{k=1}^{M-1} \alpha_k$ and $\gamma_{Mj} = -\sum_{k=1}^{M-1}\gamma_{kj}$ are included. Moreover, for the identification problem, $\alpha_0 = 0$ is set in this paper. Thus, the AIDS model leads to the nonlinear demand system, because it includes

$(\gamma_{ij} - \beta_i \alpha_j)$ and $\beta_i \gamma_{kj}$. Thus, for the AIDS model, the first $M - 1$ equations in (2.5) are simultaneously estimated as the nonlinear demand system.

As for the LA-AIDS model, the price index data is constructed by Equation (2.3), which does not depend on parameters α_i, β_i, and γ_{ij}. Therefore, we can simply estimate the first $M - 1$ equations in (2.1) as the linear demand system.

2.2.2. Elasticity

Let $\eta_{i,t}^E$ be the expenditure elasticity of the ith good at time t and $\eta_{ij,t}^P$ be the price elasticity of the jth good in the ith equation at time t. According to Alston et al. (1994), based on Equations (2.1) and (2.2), the expenditure and price elasticities of the AIDS model are obtained as follows:

$$\eta_{i,t}^E = 1 + \frac{\beta_i}{w_{it}}, \tag{2.6}$$

$$\eta_{ij,t}^P = -\delta_{ij} + \frac{\gamma_{ij}}{w_{it}} - \frac{\beta_i \alpha_j}{w_{it}} - \frac{\beta_i}{w_{it}} \sum_{k=1}^{M} \gamma_{kj} \ln p_{k,t}, \tag{2.7}$$

where $\delta_{i,j} = 1$ if $i = j$, and $\delta_{i,j} = 0$ otherwise. From the parameter estimates, $\eta_{i,t}^E$ and $\eta_{ij,t}^P$ are computed based on Equations (2.6) and (2.7).

According to Green and Alston (1990), when we use the Stone price index (2.3) in (2.1), i.e., the LA-AIDS model, the expenditure and price elasticities are given in the following forms, respectively:

$$\eta_{i,t}^E = 1 + \frac{\beta_i}{w_{i,t}}(1 - \sum_{j=1}^M w_{j,t}(\eta_{j,t}^E - 1)lnp_{j,t}), \qquad (2.8)$$

$$\eta_{ij,t}^P = -\delta_{ij} + \frac{\gamma_{ij}}{w_{i,t}} - \frac{\beta_i\alpha_j}{w_{i,t}}(w_{j,t} + \sum_{k=1}^M w_{k,t}(\eta_{kj,t}^P + \delta_{kj})lnp_{k,t}),$$

$$(2.9)$$

where $\eta_{i,t}^E$ and $\eta_{ij,t}^P$ are solved as simultaneous equations.

2.2.3. Problems in the AIDS Model

A number of studies have discussed the relationship between the AIDS and LA-AIDS. To avoid the criticisms of the LA-AIDS model that uses the Stone price index (e.g., see Buse 1994, 1998; Moschini 1995; Feenstra and Reinsdorf 2000, for the criticisms), recent empirical studies tend to shift gradually from the LA-AIDS model to the AIDS model. Nevertheless, the AIDS model has some problems, which are discussed in Tanizaki and Mizobuchi (2014) as below.

First, an assumption of the normally distributed error is imposed in some studies. The dependent variable $w_{i,t}$ in the AIDS model (2.1) lies on the interval between zero and one. Numerous studies use the FIML method, which assumes a normal distribution on the disturbance term $u_{i,t}$ (e.g., see Rossi 1988; Oladosu 2003). When a disturbance term is assumed to be normal, a dependent variable should take the range from $-\infty$ to ∞. Moreover, some studies utilize the Bayesian estimation method, where the normal distribution is assumed for the error $u_{i,t}$ (e.g., see Tiffin and Aguiar 1995; Xiao et al. 2007).

Second, in previous studies, there is no statistical inference on the expenditure and price elasticity estimates. In the empirical studies on demand analysis, usually, readers are interested in the estimates of elasticities $\eta_{i,t}^E$ and $\eta_{ij,t}^P$ in (2.6 – 2.9), not in the estimates of parameters α_i, β_i, and γ_{ij} in the AIDS model (2.1). Most of the previous empirical studies show only point estimates of the elasticities, but do not present other statistical information, such as the standard error (SE), p-value, and confidence interval. There are a few studies which show SE (e.g., see Pashardes 1993), where the SE is usually based on the Delta method. Freedman and Peters (1984) conclude that the SE based on the Delta method is underestimated in general and that the SE which utilizes the bootstrap procedure might be more plausible.

Third, there is an endogeneity problem in the AIDS model (2.1). Both the total expenditures, denoted by $X_t = \sum_{i=1}^{M} X_{i,t}$, and the expenditure share of the ith good, $i = 1$ denoted by $w_{i,t} = X_{i,t}/X_t$, are expressed as a function of $X_{i,t}$. Therefore, there is a correlation between the disturbance term $u_{i,t}$ and the explanatory variable X_t. We have the biased estimates of α_i, β_i, and γ_{ij} in the AIDS model (2.1) when we estimate the AIDS model by SUR. Numerous studies utilize SUR or the Bayesian estimation method, ignoring the endogeneity problem. Recently, however, taking into account endogeneity, the studies applying the Generalized Method of Moments (hereafter, GMM) to the AIDS model have gradually increased.

Thus, the AIDS model involves some serious problems in the empirical studies. For the first and second problems, Tanizaki and Mizobuchi (2014) considered adopting the bootstrap procedure for estimation, where any distribution is not assumed for the error term. Using the bootstrap sample based on the originally observed data, the expenditure and price elasticity estimates can be computed. Moreover, the standard errors (SE), p-values, and confidence intervals of the elasticities can be obtained. Freedman and Peters (1984), Green et al. (1987), and Krinsky and Robb (1991) have suggested the bootstrap procedure to obtain the standard errors of the parameter estimates. In the case of the AIDS model, the dependent variable $w_{i,t}$ has to lie on the interval between zero and one. Therefore, Tanizaki and Mizobuchi (2014) utilized the pairs bootstrap, rather than the residual bootstrap. We show these method in the next section. Moreover, as a solution of the third problem (i.e., the endogeneity problem), they utilize the 3SLS estimation, which yields consistent parameter estimates in the AIDS model.

2.3. MBB AND PB METHODS

Tanizaki and Mizobuchi (2014) applied the bootstrap method to obtain empirical distributions for expenditure and price elasticity estimators. Moreover, they combined the bootstrap method with the MBB method (e.g., see Künsch 1989; Liu and Singh 1992; Fitzenberger 1997) for taking account of serially correlated errors. This section introduces their method.

Assuming that all the data are I (1) processes and that there is a co-integration relationship in (2.1), Tiffin and Balcombe (2005) apply the bootstrap procedure to the LA-AIDS model and show the testing procedure on symmetry and homogeneity, where the FM-SUR method is applied for estimation. However, their analysis is based on the misleading assumption, i.e., $w_{i,t} = X_{i,t}/X_t \sim I$ (1), where they do not take into account the constraint that the share $w_{i,t}$ lies on the interval between zero and one.

Taking into account the constraint that $w_{i,t}$ lies on the interval between zero and one for all $i = 1, 2,..., M$ and $t = 1, 2,..., T,$ the PB method is adopted (e.g., see Efron and Tibshirani 1993; and MacKinnon 2006 for the PB method). In the PB method, both dependent and independent data observed at time t are always chosen in pairs. Furthermore, to avoid the endogenous problem in the AIDS model, the 3SLS is applied for estimation.

The MBB and PB methods are implemented as follows. Let us define the observed data vector at time t as $y_t = (w_t, p_t, X_t)$ for $t=1$, 2,..., T, where $w_{i,t} = (w_{1,t}, w_{2,t}, ..., w_{M,t})$ and $p_t = (p_{1,t}, p_{2,t}, ..., p_{M,t})$, i.e., w_t and p_t are $1 \times M$ vectors and accordingly y_t is a $1 \times (2M+1)$ vector. When the error terms are serially correlated, it is known that the MBB method is an efficient tool (e.g., see Künsch, 1989; Liu and Singh, 1992; Fitzenberger, 1997). Suppose that T and b are positive integers. Let b be the number of the date vectors included in one block. Then, we can construct $T - b + 1$ blocks of the observed data vectors as follows:

$$Y_{t,b} = \begin{pmatrix} y_t \\ y_{t+1} \\ \vdots \\ y_{t+b-1} \end{pmatrix}, \quad \text{for } t = 1, 2, \dots, T - b + 1,$$

where $Y_{t,b}$ denotes a $b \times (2M + 1)$ matrix. Let m^* be the maximum integer of T/b. We take $m = m^*$ when the remainder of T/b is zero and $m = m^* + 1$ otherwise.

The MBB and PB methods resample m blocks randomly with replacement out of $T - b + 1$ overlapping blocks, i.e., $\{Y_{1,b}, Y_{2,b}, \dots, Y_{T-b+1,b}\}$. Let $\{I_1, I_2, \dots, I_m\}$ be random numbers generated from the discrete uniform distribution on $\{0, 1, \dots, T - b\}$. The MBB pseudo data y_t^*, $t = 1, 2, ..., T$, is the result of arranging the elements of the m resampled blocks $\{Y_{I_1+1,b}, Y_{I_2+1,b}, \dots, Y_{I_m+1,b}\cdot\}$, which can be represented as follows:

The 1st Block   The 2nd Block   ...,   The mth Block

$(i.e., Y_{I_1+1,b})$ $(i.e., Y_{I_2+1,b})$ $(i.e., Y_{I_m+1,b^*})$

$$y_1^* = y_{I_1+1},\qquad y_{b+1}^* = y_{I_2+1}, \quad ..., \quad y_{(m-1)b+1}^* = y_{I_{m-1}+1},$$
$$y_2^* = y_{I_1+2},\qquad y_{b+2}^* = y_{I_2+2}, \quad ..., \quad y_{(m-1)b+2}^* = y_{I_{m-1}+2},$$
$$...,\qquad\qquad ...,\qquad\qquad ...,\qquad\qquad ...,$$
$$y_b^* = y_{I_1+b},\qquad y_{2b}^* = y_{I_2+b}, \quad ..., \quad y_{(m-1)b+b^*}^* = y_{I_{m-1}+b^*},$$

where the integer b^* is chosen from $T = (m-1)b + b^*$ for $b^* \leq b$. That is, Y_{I_m+1,b^*} indicates the first b^* row vectors out of $Y_{I_m+1,b}$, taking into account the end block. Thus, $y_t^* = (w_t^*, p_t^*, X_t^*)$ represents the tth row vector of the MBB and PB samples. Replacing y_t by y_t^* in (2.1)–(2.3), we estimate $(\alpha_i, \beta_i, \gamma_{ij})$ by 3SLS, where (2.2) for AIDS or (2.3) for LA-AIDS is taken. Because Tiffin and Balcombe (2005) do not consider this problem, there might be a possibility of overestimating the standard errors of the parameter estimates.

The bootstrap procedure shown above is repeated n times. From the n sets of the parameter estimates, n elasticities are computed by (2.6) and (2.7). Based on the n elasticity estimates, we have the empirical distributions of the expenditure and price elasticities. Thus, the standard errors, p-values, and confidence intervals are obtained from the empirical distributions. In the empirical analysis of the next section, we set $n = 10{,}000$, $M = 10$, and $T = 512, 180, 255, 77$. Thus, MBB and PB are combined to solve the serially correlated errors and the limited dependent variables. In the PB method, both dependent and independent data observed at time t are always selected in pairs. Therefore, the PB method does not require the assumption that the error terms are homoscedastic. However, a point to be noted is that we cannot use the lagged variables for the instrumental variables, because the order of the originally observed data are resampled.

2.4. ESTIMATION OF THE AIDS MODEL

In this section, we will estimate the AIDS model using the method in the previous section, and verify the magnitude of the price elasticity of household energy demand.

2.4.1. Data

For estimating the AIDS model, we need household expenditure data. The *National Survey of Family Income and Expenditure* (Statistical Survey Department, Statistics Bureau, Ministry of Internal Affairs and Communications) reported monthly Japanese household expenditure data, which are classified into ten categories as follows:

- Food,
- Housing,
- Electricity, gas, and water charges (hereafter, Fuel),
- Furniture and household utensils (hereafter, Furni),
- Clothes and footwear (hereafter, Clothes),
- Medical care (hereafter, Medical),
- Transport and communication (hereafter, Trans),
- Education (hereafter, Edu),
- Culture and recreation (hereafter, Culture), and
- Other consumption expenditures (hereafter, Other).

All the expenditure data are seasonally adjusted. The price data are taken from the Consumption Price Index (CPI). All the price data are divided by seasonally adjusted General CPI data before the logarithmic transformation. Thus, we use relative prices, where the base year is 2015. Total expenditure X_t is constructed by summing up all the seasonally adjusted expenditure data divided by the corresponding seasonally adjusted price data. The data period is from January 1975 to August 2017.

2.4.2. Estimation Results

As we mentioned in Section 2.2.3, the AIDS model has an endogeneity problem, and accordingly, we cannot obtain consistent parameter estimates using the classical estimation methods such as SUR. Therefore, the AIDS model (2.1) is estimated using 3SLS, where the log relative prices $(lnp_{1,t}, lnp_{2,t}, ..., lnp_{M,t})$, their cross-terms and the constant term are used for the instrumental variables. The first $M-1$ equations are simultaneously estimated (the Other consumption expenditure equation, i.e., the Mth equation, is omitted from estimation because of the additivity constraints), where $M = 10$ is taken. This paper uses the MBB and PB methods as shown in Section 2.3, where this paper takes $b = 12$ for the length of block ($b = 12$ implies that one block consists of one year data). We generate n bootstrap samples of y_t^*, and estimate α_i, β_i and γ_{ij} for each bootstrap sample, where $n = 10,000$ is set.

The demand structure of Japanese households has been possibly changed during the bubble economy period, that is, the last half of the 1980s, and the Great East Japan Earthquake disaster in March 2011, respectively. Therefore, we divide the samples into three regimes. The first regime is from January 1975 to December 1989; the second regime is from January 1990 to March 2011; and the third regime is from April 2011 to August 2017.[4] Table 2.1 shows the number of significant parameter estimates at 5% significance level for each estimation period: (i) full sample, which is denoted by 75–17 in Table 2.1, where the estimation period is from January 1975 to August 2017, and accordingly, the sample size is $T = 512$; (ii) the first regime, denoted by 75–89, where the estimation period is from January 1975 to December 1989 and the sample size is $T = 180$; and (iii) the second regime, denoted by 90–11, where the estimation period is from January 1990 to March 2011 and the sample size is $T = 255$; and (iv) the third regime, denoted by 11–17, where the estimation period is from April 2011 to August 2017 and the sample size is $T = 77$. In Table 2.1, 3SLS indicates the 3SLS estimation results using the originally observed data, while Bootstrap+3SLS represents 3SLS with the MBB and PB methods, which has been discussed in Section 2.3. In the case of Bootstrap+3SLS, each value indicates the number of parameter estimates in which the p-value is less than 0.025 or larger than 0.975. Note that in this paper, the p-value is obtained by dividing the number of positive coefficient estimates by the number of bootstrap samples,

---

[4] Actually, we tested the structural changes between 1989 and 1990, and between 2011 and 2012 as follows. Let $\hat{\theta}_1$ and $\hat{\theta}_2$ be the parameter vectors in the first and second regimes, respectively. $\hat{\theta}_1$ and $\hat{\theta}_2$ consist of $(\alpha_i, \beta_i, \gamma_{ij})$ for $i = 1, 2, ..., M - 1$ and $j = 1, 2, ..., M$. From the n bootstrap samples of y_t^*, $t = 1, 2, ..., T$, we can compute n estimates of the parameter vector. Let $\hat{\bar{\theta}}_1$ and $\hat{\bar{\theta}}_2$ be the sample averages obtained from the n estimates of parameter vectors, where the subscript denotes the first and the second regimes. $\hat{\Sigma}_1$ and $\hat{\Sigma}_2$ represent the sample variance-covariance matrices, which are also computed from the n estimates of parameter vectors in the first and second regimes. Under the null hypothesis $H_0 : \theta_1 = \theta_2$, asymptotically, we can test the structural change as follows:

$$(\hat{\bar{\theta}}_1 - \hat{\bar{\theta}}_2)'(\hat{\Sigma}_1 + \hat{\Sigma}_2)^{-1}(\hat{\bar{\theta}}_1 - \hat{\bar{\theta}}_2) \to \chi^2(108).$$

Note that we have 120 parameters in (2.1) but there are 12 additivity constraints. Therefore, the degree of freedom is given by $120 - 12 = 108$. We obtained 415.9 for the above test statistic, and the critical values of 5% and 1% are given by $\chi^2_{0.05}(108) = 133.3$ and $\chi^2_{0.01}(108) = 145.1$, respectively. Thus, we rejected the null hypothesis which states that there are no structural changes between 1989 and 1990, and between 2011 and 2012.

Moreover, we examined the tests of structural change for the individual parameters. As a result, the null hypotheses are rejected for 69 parameters out of 108 parameters between 1989 and 1990, and are rejected for 87 parameters out of 108 parameters f at 5% significance level. Therefore, we can conclude that structural changes have occurred in both the periods.

which corresponds to the empirical probability, which is larger than zero. Comparing 3SLS with Bootstrap+3SLS, we find that 3SLS is larger than Bootstrap+3SLS in ratio for all the cases, which implies that the standard errors of coefficients estimated by 3SLS are underestimated. The numbers of significant parameters in 75–17 are larger than those in 75–89, 90–11, and 11–17 for both 3SLS and Bootstrap + 3SLS.

Table 2.1. The number of significant parameters

Estimation Period	3SLS			Bootstrap+3SLS		
	α_i	β_i	γ_{ij}	α_i	β_i	γ_{ij}
75–17	7 (9)	7 (9)	55 (90)	7 (10)	5 (10)	30 (100)
75–89	3 (9)	3 (9)	21 (90)	3 (10)	3 (10)	19 (100)
90–11	3 (9)	3 (9)	23 (90)	4 (10)	4 (10)	15 (100)
11–17	4 (9)	4 (9)	20 (90)	3 (10)	3 (10)	14 (100)

2.4.3. Serial Correlation

We check whether the assumption on the uncorrelated disturbance over time t is plausible. Figure 2.1 shows the residual plots estimated by 3SLS for the first $M - 1$ equations in the AIDS model given by Equations (2.1) and (2.2), where we consider the structural changes at the beginning of 1990 and 2011, and accordingly, the estimation period is divided into three regimes ($M = 10$ is taken in this paper). The vertical lines between December 1989 and January 1990, and between March 2011 and April 2012 represent the time period when the bubble economy burst and the Great East Japan Earthquake occurred, respectively. Moreover, Figure 2.1 shows the Durbin–Watson (*DW*) statistics for each regime. The *DW* on the left-hand side corresponds to the first regime, that in the middle-hand side indicates the second regime, and that in the right-hand side indicates the third regime. Utilizing the TSP source code in the website: https://web.stanford.edu/~clint/bench/dwcrit.htm, the *DW* lower and upper bounds for the 5% critical value, denoted by *dl* and *du*, are given by (*dl*, *du*) = (1.628, 1.887) for (*T*, *k*) = (180,12), (*dl*, *du*) = (1.706, 1.885) for (*T*, *k*) = (255,12) and (*dl*, *du*) = (1.321, 1.965) for (*T*, *k*) = (77,12), where *T* denotes the sample size and *k* indicates the number of regressors including the intercept.

We take $k = M + 2 = 12$, $T = 180$ in the first regime, $T = 255$ in the second regime, and T = 77 in the third regime. The serial correlation in the error term is found for Food, Fuel, and Culture in the first regime; for Food, Fuel, Edu, and Culture in the second regime; and for Fuel in the third regime (see × in the superscript of *DW* in Figure 2.1). In the first regime, for Clothes, Trans, and Edu; in the second regime, for Furni, Clothes, and Trans; and in the third regime, for Food, Housing, Furni, Clothes, Medical, Trans, and Culture, we cannot judge whether there is a serial correlation in the error term (see △ in the superscript of *DW* in Figure 2.1). Six out of 27 cases result in no serial correlation. Here, we can confirm that energy expenditure causes structural changes twice.

Although *DW* is an approximate measure of serial correlation because the AIDS model is nonlinear in parameters, it might be plausible to utilize the MBB method, taking into account serial correlation. In addition, from Figure 2.1, we can observe heteroscedastic errors for some of the 27 cases. The PB method is useful for the heteroscedastic error terms because of resampling in pairs. The simultaneous use of MBB and PB methods solves both serial correlation and heteroscedasticity problems.

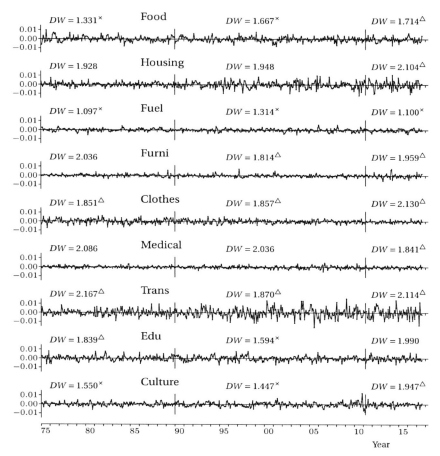

$\times$ in the superscript of DW implies that the error terms are serially correlated at 5% significance level, while the triangle indicates that we cannot judge whether there is a serial correlation.

Figure 2.1. Residual plots.

2.4.4. Tests of Homogeneity, Symmetry, and Negativity Conditions

This subsection tests the homogeneity, symmetry, and negativity conditions, which have been discussed in Section 2.2.1.

2.4.4.1. Test of Homogeneity

The first four columns of Table 2.2 show the testing results on homogeneity by the Wald test statistics, where 3SLS is used for estimation. The null hypothesis is written as follows:

$$H_0: \sum_{j=1}^{M} \gamma_{ij} = 0, \qquad \text{for } 1,2,\ldots,M$$

In Table 2.2, * indicates that the null hypothesis is rejected at the 5% significance level (note that the 5% critical value is $\chi_{0.05}^2(1) = 3.841$. For the estimation period from January 1975 to August 2017, Food, Housing, Fuel, and Trans expenditures satisfy the homogeneity condition, while the other six expenditures do not.

Table 2.2. Test of homogeneity

	(a) 3SLS (Wald test statistic)				(b) Bootstrap+3SLS (p-values)			
	75–17	75–89	90–11	11–17	75–17	75–89	90–11	11–17
Food	3.009	0.209	2.828	0.013	.6899	.2973	.0627	.4844
Housing	0.086	2.548	1.796	0.844	.5953	.0323	.9695	.8409
Fuel	0.160	3.310	4.208*	1.364	.4358	.0259	.0366	.2658
Furni	12.83*	1.390	0.747	0.922	.9005	.8894	.8317	.7417
Clothes	15.06*	3.741	0.545	6.425*	.0617	.9815*	.5463	.0252
Medical	54.58*	0.069	0.006	1.743	.0001*	.3867	.3641	.0309
Trans	0.007	0.099	0.099	0.552	.4255	.4816	.8410	.7361
Edu	7.369*	4.388*	4.944*	0.558	.9911*	.9914*	.7892	.7083
Culture	7.889*	0.438	0.115	5.686*	.2361	.2104	.3324	.0281
Other	—	—	—	—	.8061	.3690	.1406	.6452

For the homogeneity test in the bootstrap procedure based on 3SLS, the test statistic is given by $\sum_{j=1}^{M} \hat{\gamma}_{ij}^*$, where $\hat{\gamma}_{ij}^*$ denotes the parameter estimate based on the bootstrap sample: $y_t^* = (w_t^*, p_t^*, X_t^*)$ for $t = 1, 2, \ldots, T$, where $w_t^* = (w_{1,t}^*, \ldots, w_{M,t}^*)$ and $p_t^* = (p_{1,t}^*, \ldots, p_{M,t}^*)$. Using n bootstrap samples, we can obtain the n test statistics. If the number of the test statistics, which are less than zero, is less than $0.025n$ or greater than $0.975n$, we reject the null hypothesis H_0. In the last four columns in Table 2.2, the bootstrap procedure based on 3SLS is utilized to obtain the p-values, which are defined as the upper empirical probability of $\text{Prob}(\sum_{j=1}^{M}\hat{\gamma}_{ij}^{*} > 0)$ in this study. In the case of the bootstrap method, Food, Housing, Fuel, Furni, Clothes, Trans, Culture, and Other expenditures satisfy the homogeneity condition, but the other two expenditures do not (see 75–17 in Table 2.2).

2.4.4.2. Test of Symmetry

The null hypothesis of the symmetry condition is as follows:

$$H_0: \gamma_{ij} - \gamma_{ji} = 0, \qquad \text{for } i > j \text{ and } i, j = 1, 2, \ldots, M.$$

Tables 2.3 and 2.4 show the results of testing the symmetry condition. This test is examined for both 3SLS and Bootstrap+3SLS cases as well as for the entire sample (January 1975–August 2017), the first regime (January 1975–December 1989), the second regime (January 1990–March 2011), and the third regime (April 2011-August 2017) cases. In the case of Bootstrap+3SLS, the testing procedure of the symmetry condition is as follows: (i) compute $\hat{\gamma}_{ij} - \hat{\gamma}_{ji}$ for n bootstrap samples, where $n = 10,000$; and (ii) test the null hypothesis using the p-value, that is, the upper empirical probability of $\text{Prob}(\hat{\gamma}_{ij} - \hat{\gamma}_{ji} > 0)$.

Rows (a) in Tables 2.3 and 2.4 show that 16 and 35 out of 45 cases satisfy the symmetry condition for 3SLS (see Table 2.3(a)) and Bootstrap+3SLS (see Table 2.4(a)), respectively. Dividing the sample, 36 and 37 cases in the estimation period 75–89 for both 3SLS and Bootstrap+3SLS (see Tables 2.3(b) and 2.4(b)), 35 and 41 in 90–11 satisfy the symmetry condition for 3SLS and Bootstrap+3SLS (see Tables 2.3(c) and 2.4(c)), and 37 and 42 in 11–17 satisfy the symmetry condition for 3SLS and Bootstrap+3SLS (see Tables 2.3(d) and 2.4(d)). Thus, Bootstrap+3SLS satisfies the symmetry condition for more cases, compared with 3SLS. From these results, it might be concluded that similar to the case of homogeneity, we can observe structural changes between December 1989 and January 1990, and between March 2011 and April 2011 because for Tables 2.3 and 2.4, (a) is smaller than (b), (c), and (d) with respect to the number of cases where the symmetry condition is satisfied. Thus, taking into account the structural change, both homogeneity and symmetry conditions are satisfied in most cases.

Table 2.3. Test of symmetry (Wald test statistics) – 3SLS

	Food	Housing	Fuel	Furni	Clothes	Medical	Trans	Edu	Culture
(a) Estimation period: 75–17									
Housing	0.763								
Fuel	1.667	5.520*							
Furni	19.41*	38.44*	0.151						
Clothes	15.13*	7.289*	0.008	3.195*					
Medical	26.72*	11.11*	0.259	0.118	9.290*				
Trans	14.79*	0.300	23.45*	0.304	29.71*	9.121*			
Edu	1.063	7.222*	9.441*	0.047	15.85*	38.43*	29.96*		
Culture	4.385*	2.487	3.399	1.842	6.286*	12.07*	0.068	20.96*	
Other	0.190	13.67*	13.40*	1.008	5.283*	5.756*	6.071*	9.383*	7.163*
(b) Estimation period: 75–89									
Housing	5.239*								
Fuel	0.908	2.604							
Furni	7.117*	0.003	0.073						
Clothes	13.40*	0.448	1.165	2.203					
Medical	0.120	1.074	1.307	0.492	6.078*				
Trans	0.000	0.092	0.000	1.437	1.802	0.691			
Edu	1.729	3.704	3.736	0.912	0.022	0.520	0.873		
Culture	0.097	0.727	10.32*	0.470	22.85*	2.308	8.919*	5.654*	
Other	0.646	0.057	0.063	5.733*	0.064	0.533	1.626	1.107	3.521
(c) Estimation period: 90–11									
Housing	5.928*								
Fuel	2.750	0.532							
Furni	0.703	0.327	0.525						
Clothes	8.134*	3.339	3.136	0.196					
Medical	7.270*	1.814	0.097	0.913	0.839				
Trans	4.598*	3.784	27.31*	0.464	0.121	0.272			
Edu	15.69*	0.076	1.829	8.797*	1.068	2.026	4.330*		
Culture	0.098	0.501	0.788	1.068	0.506	1.779	2.800	1.992	
Other	0.076	0.261	5.673*	1.101	0.190	0.330	4.449*	3.737	0.445
(d) Estimation period: 11–17									
Housing	0.055								
Fuel	0.019	0.044							
Furni	10.04*	1.167	1.838						
Clothes	0.000	4.631*	0.030	0.126					
Medical	0.713	0.432	0.130	0.084	0.097				
Trans	4.214*	0.266	6.597*	0.526	3.183	2.291			
Edu	9.836*	1.947	0.039	0.854	2.143	0.734	3.631		
Culture	0.826	5.573*	9.518*	0.039	0.643	1.457	0.198	3.571	
Other	0.445	1.225	5.918*	0.244	0.198	0.066	0.004	0.403	0.512

Table 2.4. Test of symmetry (*p*-values) – Bootstrap + 3SLS

	Food	Housing	Fuel	Furni	Clothes	Medical	Trans	Edu	Culture
(a) Estimation period: 75–17									
Housing	.7252								
Fuel	.3321	.7000							
Furni	.8065	.9968*	.9626						
Clothes	.0397	.0675	.4810	.8303					
Medical	.0014*	.2850	.0112*	.0006*	.7224				
Trans	.4042	.5009	.9067	.0004*	.0703	.9980*			
Edu	.9994*	.9848*	.2731	.9973*	.9874*	.6475	.4141		
Culture	.2935	.0652	.1205	.8690	.9639	.9667	.1101	.8269	
Other	.4351	.6633	.9493	.9074	.6345	.9325	.8886	.0375	.8569
(b) Estimation period: 75–89									
Housing	.0018*								
Fuel	.2513	.1025							
Furni	.4779	.6251	.8472						
Clothes	.9914*	.9083	.9534	.5039					
Medical	.4003	.8163	.5553	.5950	.0107*				
Trans	.4848	.2011	.7408	.6925	.3971	.0481			
Edu	.9999*	.9946*	.4183	.2838	.9903*	.9680	.0118*		
Culture	.6310	.5616	.4886	.2903	.2954	.7991	.3067	.8832	
Other	.2060	.4284	.8342	.1475	.1904	.2433	.6991	.0478	.9680
(c) Estimation period: 90–11									
Housing	.9852*								
Fuel	.1110	.3553							
Furni	.2731	.8696	.9993*						
Clothes	.5512	.1822	.5726	.9648					
Medical	.3026	.2956	.3579	.0809	.9338				
Trans	.8996	.7658	.8396	.1913	.3507	.9841*			
Edu	.9532	.5230	.8660	.4539	.4847	.3871	.4590		
Culture	.7200	.0037*	.3805	.5491	.8911	.5357	.3785	.2471	
Other	.1008	.1316	.9025	.6265	.0995	.0506	.2330	.1767	.3325
(d) Estimation period: 11–17									
Housing	.4501								
Fuel	.5805	.4026							
Furni	.1823	.5211	.9907*						
Clothes	.0247*	.1837	.3722	.8395					
Medical	.9351	.6078	.1445	.9231	.9558				
Trans	.8580	.2239	.8739	.1076	.5971	.9371			
Edu	.1879	.4786	.9688	.0710	.7363	.7136	.8840		
Culture	.0085*	.0580	.1487	.1734	.1548	.7297	.3302	.7873	
Other	.7494	.3234	.6602	.0441	.7165	.9721	.3252	.1398	.8458

2.4.4.3. Test of Negativity

Finally, we conduct the test of negativity (or concavity) condition, which shows whether the estimated parameters guarantee the condition of local maximization. For each bootstrap sample, the negativity can be checked by calculating the eigenvalues of the Slutsky matrix (or equivalently, the substitution matrix, denoted by S), where an element in the ith row and jth column is denoted by s_{ij}. The negativity condition implies that S is a negative semidefinite matrix, that is, all the eigenvalues are non-positive. Deaton and Muellbauer (1980) suggest to use C, whose eigenvalues has the same signs as those of S, where the (i, j)th element of C is given by the following:

$$C_{ij} = \gamma_{ij} + \beta_i\beta_j \ln\frac{X_t}{P_t} - w_{i,t}\delta_{ij} + w_{i,t}w_{j,t},$$

where δ_{ij} is the Kronecker delta, that is, $\delta_{ij} = 1$ for $i = j$ and $\delta_{ij} = 0$ for $i = j$.[5] This test is examined for the full sample (January 1975–August 2017), the first regime (January 1975–December 1989), the second regime (January 1990–March 2011), and the third regime (April 2011–August 2017) in Bootstrap+3SLS. The negativity condition is satisfied in only 15 cases for the full sample, in 386 cases for the first regime, in two cases for the second regime, and in one case for the third regime, out of 10,000 bootstrap parameter estimates. That is, we can conclude that the negativity condition is rejected.[6] It is extremely difficult to obtain the parameter estimates which satisfy the negativity condition.

2.5. PRICE AND EXPENDITURE ELASTICITIES

Tables 2.5 and 2.6 show the expenditure and own price elasticities based on Equations (2.6) and (2.7) in the case of (a) the full sample (January 1975– August 2017), (b) the first regime (January 1975–

---

[5] Because c_{ij} depends on $w_{i,t}$, $\ln X_t$ and $\ln P_t$ we evaluate them at the sample averages of $w_{i,t}$, $\ln X_t$, and $\ln P_t$ over time t. That is, $w_{i,t}$, $\ln X_t$ and $\ln P_t$ in c_{ij} are replaced by $\bar{w}_i$, $\overline{\ln X}$ and $\overline{\ln P}$ (i.e., sample means over t), respectively.
[6] Moschini (1998) suggests to use the semiflexible AIDS model, which imposes the negativity condition.

December 1989), (c) the second regime (January 1990–March 2011), and the third regime (April 2011–August 2017).[7] As discussed in Section 2.3, using the bootstrap procedure, we can obtain the expenditure and price elasticities, the standard error, the *p*-value and the 2.5 and 97.5 percent points (i.e., 95 % confidence interval), which are denoted by Elas, SE, *p*-value, 2.5 and 97.5%, respectively. For comparison, in Tables 2.5 and 2.6, we also show each elasticity (Elas) and its standard error (SE) in the case where 3SLS is utilized. SE in 3SLS is given by the Delta method.[8]

2.5.1. Expenditure Elasticity

From Bootstrap + 3SLS in Table 2.5 (i.e., the expenditure elasticities), eight or nine out of 10 expenditures are statistically different from zero for each estimation period. A good is elastic, which is known as a luxury good, when the expenditure elasticity is larger than 1, and a good is inelastic, which is known as a necessity good, when the expenditure elasticity is smaller than 1. In Table 2.5 (a), Furni, Clothes, Trans, Edu, Culture, and Other are luxury goods, and Food, Housing, Fuel, and Medical are necessity goods. Moreover, a good is known to be inferior when the expenditure elasticity is smaller than 0. The expenditure elasticity of Fuel (January 1975–December 1989) is negative but statistically insignificant. This result is not

---

[7] As it is the case of the negativity test, in Tables 2.5 and 2.6, we evaluate $w_{i,t}$ and $lnp_{i,t}$ at the sample averages of $w_{i,t}$ and $lnp_{i,t}$ over time t. Remember that the expenditure and price elasticities are given by (2.6) and (2.7), which depend on $w_{i,t}$ and $lnp_{i,t}$.

[8] The Delta method is as follows. Suppose that an elasticity η is expressed as a function $\eta(\theta)$ of a parameter vector $\theta = (\alpha_i, \beta_i, \gamma_{ij})$ for $i = 1,2,..., M - 1$ and $j = 1,2,..., M$. Let V be the variance–covariance matrix of $\hat{\theta}$ and D be the gradient of $\eta(\theta)$. The asymptotic standard errors, denoted by SE(η), are obtained as follows:
$$SE(\eta) \approx D'VD.$$
Here, because of the additivity constraint, the Mth expenditure parameters are represented by those of the other expenditure parameters. Therefore, it is not easy to obtain the standard errors of the Mth elasticities.

plausible because Fuel is known more commonly as the normal good (that is, the expenditure elasticity should be greater than 0). Here, the expenditure elasticity of the fuel is not statistically significant except in the third period (April 2011–August 2017). As we mentioned in Section 2.1, due to power shortage after the Great East Japan Earthquake in March 2011, a large-scale power-saving request was issued in various places, and electricity supply and demand attracted public attention. Since the Great East Japan Earthquake, people may be interested in energy demand (especially, electricity demand), which may have caused a change in the size of elasticity.

2.5.2. Own Price Elasticity

Table 2.6 shows the uncompensated own price elasticity given by Equation (2.7). The results in the full sample case (Table 2.6(a)) indicate that all the own price elasticities except for Housing are statistically significant and negative. Therefore, the demand principle holds in the nine expenditure items. However, the elasticity of Housing is positive but statistically insignificant. Taking into account the structural change, the elasticity of Housing is negative and significant in the first regime (Table 2.6(b)), positive and significant in the second regime (Table 2.6(c)), and positive but insignificant in the third regime (Table 2.6(d)). Therefore, it might be concluded that for Housing, the demand principle holds only in the first regime. For Food, Fuel, Furni, Clothes, and Culture, the price elasticities in the second regime are larger in absolute value than those in the first regime. It might be plausible to consider that the long-run deflation in the second regime forces Japanese households to be sensitive against the price increase. Here, we confirm whether the parameter shifts of some elasticities occurred during the bubble economy period.

Meanwhile, remarkable changes can be confirmed even after the Great East Japan Earthquake. From Table 2.6(d), except for Fuel, Clothes, Medical, Culture, these changes are not statistically significant. Furthermore, not only for Housing, but also the sign of Furni is not significant, but it has become positive. The magnitude of price elasticity is larger than other periods in many cases. From these results, it can be seen that the response to price changes of people's consumer goods has caused a structural change after the disaster. However, unlike the case of expenditure elasticity, Fuel's price elasticity could not be confirmed.

2.5.3. Testing Structural Change in the Elasticities

We have tested the structural change on individual parameters. Here, we consider testing whether there is a structural change on the elasticities. Let $\hat{\eta}_1$ and $\hat{\eta}_2$ be the sample averages from the n elasticities in the first and second regimes, respectively. $\hat{\sigma}_1$ and $\hat{\sigma}_2$ represent the standard errors of $\hat{\eta}_1$ and $\hat{\eta}_2$, which are also computed from the n elasticities. Under the null assumption that η_1 is equal to η_2, as the number of observations increase for both the first and second regimes, we have the following:

$$\frac{\hat{\eta}_1 - \hat{\eta}_2}{\sqrt{\hat{\sigma}_1^2 + \hat{\sigma}_2^2}} \to N(0,1).$$

Note that $\hat{\eta}_1$ and $\hat{\sigma}_1$ correspond to elasticity and SE in Tables 2.5 (b and c) and 2.6 (b and c), respectively, while $\hat{\eta}_2$ and $\hat{\sigma}_2$ correspond to elasticity and SE in Tables 2.5 (c and d) and 2.6 (c and d).

Table 2.5. Estimation results of expenditure elasticity

	Bootstrap+3SLS					3SLS	
	Elas	SE	*p*-value	2.5%	97.5%	Elas	SE
(a) Estimation period: 75–17							
Food	0.4844*	0.0890	1.0000	0.303	0.658	0.4837*	0.0375
Housing	0.4675	0.3188	0.9231	−0.185	1.072	0.3230	0.2107
Fuel	0.6407*	0.2025	0.9988	0.264	1.060	0.7190*	0.2145
Furni	1.6642*	0.5028	1.0000	0.857	2.794	1.3964*	0.2114
Clothes	1.5658*	0.1806	1.0000	1.188	1.908	1.5563*	0.1318
Medical	−0.1359	0.2040	0.2455	−0.545	0.256	−0.2893	0.3855
Trans	1.0615*	0.2035	1.0000	0.668	1.470	0.9535*	0.1405
Edu	1.9401*	0.3102	1.0000	1.335	2.551	2.0890*	0.2679
Culture	1.1771*	0.2151	1.0000	0.807	1.657	1.1652*	0.0873
Other	1.4140*	0.1085	1.0000	1.193	1.618	—	
(b) Estimation period: 75–89							
Food	0.4703*	0.1156	0.9999	0.237	0.689	0.6166*	0.1122
Housing	1.1616*	0.5425	0.9922	0.193	2.284	0.9281	0.6788
Fuel	−0.0292	0.3554	0.4907	−0.774	0.627	−0.7713*	0.2626
Furni	1.8703*	0.7113	0.9980	0.537	3.323	1.4820	1.3918
Clothes	1.9840*	0.4640	1.0000	1.093	2.903	2.3030*	0.1294
Medical	1.2627*	0.4497	0.9963	0.370	2.135	1.9318	0.9872
Trans	0.8771	0.5218	0.9542	−0.153	1.894	0.5624	0.5219
Edu	1.5529*	0.6291	0.9943	0.342	2.796	1.2162*	0.2900
Culture	1.1141*	0.2684	1.0000	0.598	1.656	1.3450*	0.2237
Other	1.2051*	0.2101	1.0000	0.803	1.616	—	
(c) Estimation period: 90–11							
Food	0.4567*	0.0950	1.0000	0.264	0.635	0.5520*	0.0647
Housing	1.7853*	0.3286	1.0000	1.151	2.453	1.3434*	0.2955
Fuel	0.1607	0.1476	0.8699	−0.119	0.465	−0.0165	0.2821
Furni	1.5577*	0.4861	1.0000	0.654	2.559	1.4073*	0.6793
Clothes	1.3773*	0.2797	1.0000	0.779	1.881	1.7630*	0.2806
Medical	0.2646	0.3834	0.7546	−0.506	0.978	0.5572	0.7338
Trans	1.5405*	0.3844	0.9999	0.803	2.313	1.1831*	0.2338
Edu	1.3725*	0.4776	0.9988	0.467	2.339	1.3744*	0.3572
Culture	1.1724*	0.2306	1.0000	0.736	1.652	1.2077*	0.1404
Other	1.1931*	0.1557	1.0000	0.878	1.495	—	
(d) Estimation period: 11–17							
Food	0.3271*	0.0728	0.9998	0.173	0.473	0.3910*	0.0950
Housing	2.0364*	0.7013	0.9991	0.756	3.795	1.9658	1.1040
Fuel	0.5532*	0.1357	0.9995	0.267	0.853	0.5482	2.3352
Furni	2.3977*	0.9866	0.9975	0.441	3.779	3.4270	2.6251
Clothes	1.3725*	0.2303	0.9999	0.861	1.836	1.3847	2.0442
Medical	0.6781	0.3872	0.9228	−0.216	1.213	0.9539	1.9265
Trans	2.2246*	0.2588	1.0000	1.727	2.783	1.8809*	0.8863
Edu	0.8861*	0.4757	0.9855	0.100	2.042	0.7172	2.7688
Culture	0.9626*	0.1916	1.0000	0.514	1.259	1.1046	1.2572
Other	0.6622*	0.1884	1.0000	0.386	1.113	—	

Kenichi Mizobuchi and Hisashi Tanizaki

Table 2.6. Estimation results of own price elasticities

	Bootstrap+3SLS					3SLS	
	Elas	SE	*p*-value	2.5%	97.5%	Elas	SE
(a) Estimation period: 75–17							
Food	−0.9342*	0.3768	0.0014	−1.820	−0.293	−0.8491	0.5462
Housing	1.1863	0.8704	0.9160	−0.525	2.913	1.2591*	0.3098
Fuel	−1.0216*	0.1695	0.0000	−1.356	−0.682	−1.0038*	0.2949
Furni	−1.5428*	0.2361	0.0000	−2.011	−1.072	−1.4791*	0.4994
Clothes	−1.3698*	0.2416	0.0000	−1.842	−0.885	−1.5303*	0.2373
Medical	−1.8839*	0.1779	0.0000	−2.232	−1.523	−1.9589*	0.3014
Trans	−1.5179*	0.2644	0.0000	−2.096	−1.018	−1.4938*	0.3680
Edu	−0.9758*	0.1654	0.0000	−1.307	−0.648	−1.0131*	0.2391
Culture	−2.9688*	0.5210	0.0000	−3.751	−1.672	−3.2372*	0.1093
Other	−1.6820*	0.0813	0.0000	−1.847	−1.528	—	
(b) Estimation period: 75–89							
Food	−1.3759*	0.2904	0.0000	−1.969	−0.833	−1.3193	1.1530
Housing	−3.1514*	1.2273	0.0017	−5.804	−0.921	−3.0177*	0.4495
Fuel	−1.4439*	0.1853	0.0000	−1.775	−1.044	−1.4820*	0.6158
Furni	−0.5498	0.8541	0.2442	−2.247	1.177	−0.1221	1.2286
Clothes	−1.5675*	0.3977	0.0001	−2.356	−0.802	−1.5445	1.0332
Medical	−1.9865*	0.3538	0.0000	−2.690	−1.318	−2.0135*	0.4538
Trans	−1.5388*	0.5694	0.0037	−2.677	−0.433	−1.6476	1.6109
Edu	−1.7651*	0.5186	0.0000	−2.989	−0.919	−1.6421*	0.3986
Culture	−1.5950*	0.4950	0.0032	−2.476	−0.512	−1.8623*	0.3480
Other	−1.8637*	0.1824	0.0000	−2.280	−1.540	—	
(c) Estimation period: 90–11							
Food	−1.6634*	0.3398	0.0000	−2.317	−0.976	−1.6151	1.1166
Housing	3.4570*	1.4287	0.9838	0.336	6.146	3.0027*	0.3403
Fuel	−1.6048*	0.2130	0.0000	−2.031	−1.186	−1.6143*	0.6047
Furni	−0.9574	0.5001	0.0514	−1.768	0.256	−1.1484	0.9284
Clothes	−1.7539*	0.3657	0.0000	−2.540	−1.066	−1.7153*	0.4368
Medical	−1.5331*	0.2794	0.0000	−2.118	−1.019	−1.3875*	0.2073
Trans	−0.5835	0.7683	0.2067	−1.969	1.112	−0.8324	0.7345
Edu	−1.6827*	1.4469	0.0060	−5.510	−0.269	−0.7479*	0.3087
Culture	−1.9148*	0.6337	0.0020	−3.111	−0.611	−1.5466*	0.1808
Other	−1.8508*	0.2945	0.0000	−2.513	−1.338	—	
(d) Estimation period: 11–17							
Food	−1.5913	0.8846	0.0321	−3.347	0.079	−1.3665	1.8727
Housing	5.5266	4.5258	0.9093	−2.450	14.90	3.9560*	1.8629
Fuel	−1.4749*	0.7008	0.0025	−3.006	−0.319	−1.5620	2.4745
Furni	2.9613	2.4170	0.8722	−1.048	8.004	2.2784	4.7896
Clothes	−4.3525*	1.3991	0.0002	−7.720	−2.171	−3.8624*	1.2646
Medical	−7.6251*	2.3035	0.0021	−12.03	−3.143	−5.6257*	1.7544
Trans	−0.5883	1.4741	0.3246	−3.385	2.446	−0.5494	1.6686
Edu	−3.0824	6.1555	0.2832	−14.08	10.19	−1.6525	1.2896
Culture	−2.5014*	1.3998	0.0094	−5.842	−0.410	−2.3353*	0.2262
Other	−1.7180	0.8229	0.0401	−2.857	0.679	—	

Table 2.7. Test of structural changes (*t* test) – Bootstrap + 3SLS

	Expenditure elasticity		Own price elasticity	
	89/90	11/11	89/90	11/11
Food	0.0909	1.0828	0.6432	−0.0761
Housing	−0.9834	−0.3242	−3.5086*	−0.4361
Fuel	−0.4935	−1.9576	0.5699	−0.1773
Furni	0.3628	−0.7637	0.4118	−1.5877
Clothes	1.1198	0.0132	0.3450	1.7970
Medical	1.6890	−0.7589	−1.0057	2.6254*
Trans	−1.0236	−1.4763	−0.9990	0.0029
Edu	0.2284	0.7216	−0.0536	0.2214
Culture	−0.1648	0.6998	0.3977	0.3818
Other	0.0459	2.1722*	−0.0372	−0.1519

Thus, we can test whether structural changes occurred between December 1989 and January 1990, and between March 2011 and April 2011, using the expenditure elasticities and the standard errors in Table 2.5 (b, c, and d). Similarly, using Table 2.6 (b, c, and d), we can examine the structural change of the own price elasticities. The test statistics are shown in Table 2.7, and they are compared with the standard normal distribution. From the Table, the own price elasticity of Housing has changed between December 1989 and January 1990 (see 89/90 in Table 2.7), while we cannot find any structural change on the other elasticities. On the other hand, the expenditure elasticity of Other and the own price elasticity of Medical have changed between March 2011 and April 2011 (see 11/11 in Table 2.7). Thus, we conclude that only in the case of Housing price elasticity, the structural change has occurred between December 1989 and January 1990, and in the case of Other expenditure elasticity and Medical own price elasticity, the structural change has occurred between March 2011 and April 2011.

Table 2.8. AIDS vs. LA-AIDS (75–17)

	Expenditure elasticity					Own price elasticity				
	Elas	SE	*p*-value	2.5%	97.5%	Elas	SE	*p*-value	2.5%	97.5%
(a) AIDS (Bootstrap+3SLS)										
Food	0.4844*	0.0890	1.0000	0.303	0.658	−0.9342*	0.3768	0.0014	−1.820	−0.293
Housing	0.4675	0.3188	0.9231	−0.185	1.072	1.1863	0.8704	0.9160	−0.525	2.913
Fuel	0.6407*	0.2025	0.9988	0.264	1.060	−1.0216*	0.1695	0.0000	−1.356	−0.682
Furni	1.6642*	0.5028	1.0000	0.857	2.794	−1.5428*	0.2361	0.0000	−2.011	−1.072
Clothes	1.5658*	0.1806	1.0000	1.188	1.908	−1.3698*	0.2416	0.0000	−1.842	−0.885
Medical	−0.1359	0.2040	0.2455	−0.545	0.256	−1.8839*	0.1779	0.0000	−2.232	−1.523
Trans	1.0615*	0.2035	1.0000	0.668	1.470	−1.5179*	0.2644	0.0000	−2.096	−1.018
Edu	1.9401*	0.3102	1.0000	1.335	2.551	−0.9758*	0.1654	0.0000	−1.307	−0.648
Culture	1.1771*	0.2151	1.0000	0.807	1.657	−2.9688*	0.5210	0.0000	−3.751	−1.672
Other	1.4140*	0.1085	1.0000	1.193	1.618	−1.6820*	0.0813	0.0000	−1.847	−1.528
(b) LA-AIDS (Bootstrap+3SLS)										
Food	0.4928*	0.0882	1.0000	0.311	0.665	−0.8771*	0.3943	0.0035	−1.816	−0.221
Housing	0.4966	0.3167	0.9356	−0.154	1.093	1.1988	0.8782	0.9172	−0.521	2.951
Fuel	0.6310*	0.1972	0.9986	0.259	1.039	−1.0147*	0.1684	0.0000	−1.348	−0.677
Furni	1.6107*	0.5028	1.0000	0.808	2.731	−1.5243*	0.2347	0.0000	−1.989	−1.054
Clothes	1.5831*	0.1792	1.0000	1.218	1.929	−1.3972*	0.2417	0.0000	−1.874	−0.915
Medical	−0.1256	0.2054	0.2649	−0.538	0.269	−1.8676*	0.1781	0.0000	−2.218	−1.504
Trans	0.9683*	0.2243	1.0000	0.526	1.414	−1.4805*	0.2679	0.0000	−2.059	−0.967
Edu	2.0167*	0.3280	1.0000	1.360	2.645	−0.9854*	0.1701	0.0000	−1.331	−0.651
Culture	1.1481*	0.2175	1.0000	0.774	1.636	−2.9754*	0.5217	0.0000	−3.756	−1.670
Other	1.4322*	0.1130	1.0000	1.205	1.646	−1.6961*	0.0801	0.0000	−1.857	−1.543
(c) AIDS (Bootstrap+SUR)										
Food	0.4473*	0.0502	1.0000	0.343	0.542	−0.9153*	0.3566	0.0013	−1.732	−0.293
Housing	0.9915*	0.1965	1.0000	0.601	1.372	0.8341	0.9320	0.8133	−0.932	2.748
Fuel	0.4868*	0.1153	1.0000	0.261	0.715	−1.0254*	0.1719	0.0000	−1.367	−0.681
Furni	1.7821*	0.5038	1.0000	0.992	2.826	−1.5824*	0.2402	0.0000	−2.057	−1.103
Clothes	1.4593*	0.1259	1.0000	1.206	1.703	−1.2977*	0.2232	0.0000	−1.738	−0.860
Medical	0.1509	0.1554	0.8371	−0.157	0.454	−1.7553*	0.1739	0.0000	−2.093	−1.404
Trans	1.4756*	0.1676	1.0000	1.145	1.805	−1.6750*	0.2732	0.0000	−2.291	−1.182
Edu	1.6216*	0.2329	1.0000	1.163	2.079	−0.9405*	0.1616	0.0000	−1.254	−0.607
Culture	1.1100*	0.1411	1.0000	0.862	1.414	−2.9857*	0.5085	0.0000	−3.752	−1.715
Other	1.2787*	0.0896	1.0000	1.097	1.443	−1.7022*	0.0842	0.0000	−1.870	−1.535
(d) LA-AIDS (Bootstrap+SUR)										
Food	0.4523*	0.0503	1.0000	0.346	0.547	−0.8707*	0.3739	0.0028	−1.729	−0.228
Housing	0.9864*	0.1906	1.0000	0.603	1.355	0.8451	0.9306	0.8167	−0.927	2.754
Fuel	0.4940*	0.1119	1.0000	0.272	0.714	−1.0151*	0.1696	0.0000	−1.353	−0.676
Furni	1.6824*	0.4981	1.0000	0.905	2.701	−1.5460*	0.2353	0.0000	−2.006	−1.072
Clothes	1.4451*	0.1250	1.0000	1.195	1.689	−1.2893*	0.2236	0.0000	−1.728	−0.847
Medical	0.1572	0.1563	0.8450	−0.152	0.461	−1.7390*	0.1744	0.0000	−2.077	−1.385
Trans	1.4083*	0.1757	1.0000	1.054	1.746	−1.6590*	0.2684	0.0000	−2.259	−1.168
Edu	1.6885*	0.2392	1.0000	1.227	2.162	−0.9438*	0.1632	0.0000	−1.262	−0.608
Culture	1.0693*	0.1410	1.0000	0.821	1.376	−2.9892*	0.5096	0.0000	−3.750	−1.702
Other	1.3105*	0.0940	1.0000	1.123	1.484	−1.7037*	0.0833	0.0000	−1.869	−1.538

2.5.4. AIDS vs. LA-AIDS

Tanizaki and Mizobuchi (2014) compared estimation results between AIDS and LA-AIDS models. In empirical studies, the LA-

AIDS model, which consists of Eqs. (2.1) and (2.3), is more often utilized, rather than the AIDS model given by Eqs. (2.1) and (2.2). In the LA-AIDS model, the Stone price index (2.3) is used for lnP_t in (2.1), while in the AIDS model, the original price index (2.2) is utilized for lnP_t. The Stone price index in Eq. (2.3) is known as an approximation of Eq. (2.2). In the case of LA-AIDS, expenditure, and price elasticities are given by Eqs. (2.8) and (2.9), which are discussed in Green and Alston (1990). As we can confirm from Figure 2.2, the Stone price index (2.3) does not approximate the original price index (2.2). Table 2.8 shows the difference between AIDS and LA-AIDS models for the full sample (January 1975–August 2017). (a) in Table 2.8 is exactly equivalent to (a) in Bootstrap+3SLS of Tables 2.5 and 2.6.

We compare (a) and (b) in Table 2.8, where (a) and (b) indicate AIDS and LA-AIDS, respectively. Both are estimated by the bootstrap method based on 3SLS. The elasticity estimates in (a) are very close to those in (b), where the differences between (a) and (b) are −0.0766 to 0.0932 for the expenditure elasticity and −0.0571 to 0.0274 for the own price elasticity. From (a) and (b) in Table 2.8, there is no evidence that the elasticity estimates in (b) are biased toward either the left-hand side or the right-hand side. In addition, SEs in (a) are very close to those in (b).

2.5.5. Endogeneity Problem (3SLS vs. SUR)

In order to check for the endogeneity problem on X_t, we compare the elasticities of 3SLS with those of SUR in Table 2.8, where (a) should be compared with (c), and (b) should be compared with (d). Usually, it is known that SUR yields biased and inconsistent estimates, but 3SLS gives us consistent estimates. As it is expected, we can find the divergence between both elasticities. That is, the elasticity estimates in SUR are very different from those in 3SLS. As for AIDS, the differences between (a) and (c) are −0.5240 to 0.3185 for the expenditure elasticities and −0.1286 to 0.3522 for the price elasticities. For LA-AIDS, the differences between (b) and (d) are −0.4898 to 0.3283 for the expenditure elasticities and −0.1286 to 0.3537 for the price elasticities. Moreover, in most of the cases, the standard errors of the elasticity estimates in SUR are smaller than those in 3SLS for both expenditure and price elasticities. That is, it is observed that the standard errors of SUR are underestimated. Thus, we see that the endogeneity problem is serious in both AIDS and LA-AIDS models. As it is easy to handle, empirical demand studies prefer the LA-AIDS model to the AIDS model. In the past, however, numerous studies have been conducted on accuracy of the Stone price index (2.3), compared with the original price index (2.2). For example, see Pashardes (1993), Buse (1994, 1998), Hahn (1994), Moschini (1995), Buse and Chan (2000), and Feenstra and Reinsdorf (2000). They show that the Stone price index is not a good approximation of the original price index, and therefore, they strongly suggest to use the original price index. Figure 2.2 shows the plots of the two price indices, which show that the divergence between two indices is very large. The original price index is more volatile than the Stone price index.

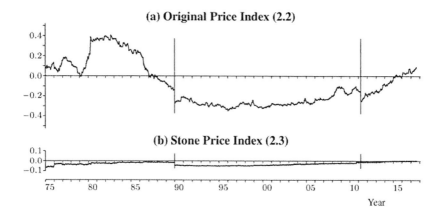

Figure 2.2. Price indices – 3SLS.

CONCLUSION

In this chapter, we examined the change in the demand structure of Japanese households owing to the Great East Japan Earthquake that occurred in March 2011 from the estimation result of the demand model. For our estimation, household expenditure data on 10 items (January 1975 to August 2017) were used and estimated by the AIDS model. Following Miobuchi and Tanizaki (2013), a method combining PB and MBB methods was used in order to consider the range of dependent variables of the demand system model and the sequence correlation of error terms. The conclusions obtained in our analysis are as follows.

First, the Great East Japan Earthquake in March 2011 revealed that there occurred a structural change in household demand. In particular, the magnitudes of expenditure elasticity of Other and price elasticity of Medical changed significantly. On the other hand, regarding energy, although the magnitude of expenditure elasticity has changed as described above, no structural change has occurred.

Second, regarding the expenditure elasticity of Fuel demand, it was estimated to be 0.47 for the entire sample period and was statistically significant at 0.55 after the disaster (third regime). Fuel is a necessity and the elasticity is less than 1; however, it was not a small value. This may be due to the fact that the number of electric appliances owned by households was high, based on the size of the income.

Third, regarding the price elasticity of Fuel demand, it was estimated to be −1.02 during the entire period, and it was slightly higher than −1.47 after the disaster. This result implies that Japanese households tend to respond significantly to changes in fuel prices. In Chapters 4 and 5 of this book, we analyze whether economic incentives are effective for power saving, but there is also a certain degree of response to incentives, which is consistent with the results of this chapter.

Finally, expenditure and price elasticities for some items were not statistically significant (these results are also consistent with Tanizaki and Mizobuchi, 2014). Therefore, it can be said that in an empirical study of elasticity using a demand system model, it is incorrect to interpret elasticity without verifying the statistical properties of the elasticity itself.

Chapter 3

THE EFFECTS OF PUBLIC PRESSURE ON POWER SAVING

ABSTRACT

Japan's nuclear power plants were shut down owing to the Great East Japan Earthquake that occurred in 2011. The growing concern about power shortage led the government to request for conservation of electricity according to regional power supply level at the time when electricity demand in summer and winter increased. In this chapter, using monthly household electricity usage fee data (panel data) from January 2003 to December 2016, we examined whether the government's energy-saving request had the effect of reducing household electricity usage. The results suggest that after the earthquake, significant power-saving effects were observed in each region. Furthermore, there was a nationwide consciousness of energy saving, as electricity saving was also carried out in areas where power-saving requests were not issued. In addition, it was confirmed that power saving has been established in some areas even after the cessation of the power-saving request since 2013.

Keywords: power-saving request, electricity saving, power shortage

3.1. INTRODUCTION

Since the accident at the Fukushima Daiichi Nuclear Power plant due to the Great East Japan Earthquake, nuclear power plants in all of Japan ceased operations due to concerns about nuclear safety. This shutdown of nuclear power plants, which accounts for about 30% of Japan's electricity supply, caused power shortage problems depending on seasons and areas. First, in the summer of 2011, a 15% energy-saving target was set up for households/areas/facilities served by the Tokyo Electric Power Company (TEPCO) and the East Northern Electric Power Company (based on the summer of 2010). Furthermore, for large power customers (contract power 500 kW or more), power usage restriction was sought.[9] In the Kansai Electric Power Company, a target of 10% or more was set, and efforts to save electricity in each division were conducted. Table 3.1 shows the power-saving targets for each year in each region of Japan. According to the Agency for Natural Resources and Energy (2011), the maximum energy-saving target of the household sector during the peak electricity demand period in 2011 were set at 11% in the TEPCO, 18% in the Tohoku Electric Power Company, and 4% in the Kansai Electric Power Company. As such, problems such as power failure due to power shortage were avoided.

By the summer of 2012, the power shortage problem spread all over Japan (except for Okinawa) due to delayed recommencement of nuclear power plants that had ceased operations due to safety inspections. While the government set the energy-saving targets in each region as of May 2012, it was concerned about severe electric power shortage, especially in the Western Japan region. Thus, the government recommended Units 3 and 4 of Ooi nuclear plant, Kansai Electric Power Company in mid-June. Consequently, there was an increase in electricity supply capacity, and the final electricity-saving targets for

---

[9] Power usage restriction was carried out over a total of 550 hours for 50 days except Saturdays, Sundays, and holidays from 1 July 1 to September 9, 2011, 11 hours from 9: 00 to 20: 00 each day.

summer 2012 were reduced to 10% in the Kansai Electric Power Company and Kyushu Electric Power Company, 5% or more in Shikoku, 7% or more in Hokkaido; in other areas, no targets were set. According to the Supply Demand Verification Committee (2012), households' energy conservation results were 10%, 12%, 8%, and 5% in the Kansai Electric Power Company, Kyushu Electric Power Company, Shikoku Electric Power Company, and the Hokkaido Electric Power Company, respectively. As a result, as of 2012, stable supply and demand conditions were maintained and power crisis was avoided.

In the summer of 2013, no targets were set because of the prospect of securing the reserve ratio of 3%, which is the standard for stable supply within all power lines. According to the Subcommittee on Electricity Supply and Demand Verification (2013), the peak electricity-saving performance was 14.7% in the Tokyo Electric Power Company and 9.8% in the Kansai Electric Power Company. In addition, according to Kansai Electric Power Company (2014), the energy-saving performance of the household sector in the pipe was reported as 9% in 2012 and 11% in 2013. As a result, in 2013, a certain amount of energy savings was achieved even without a target, which is expected to be part of firms' and households' conservation behavior so far.

In the summer of 2014, there were seldom nuclear power plant operations; however, similar to 2013, no targets were set in any area, and a request for power saving not accompanied by targets was made. According to the Kansai Electric Power Company (2014), about 13% of electricity savings had been achieved as a result of energy conservation in the household sector.

As described above, the power shortage problem that occurred since the Great East Japan Earthquake, which could have caused power crises such as large-scale blackouts, were avoided by undertaking measures requiring power-saving requests. Even today, as of December 2017, most nuclear power plants have ceased to operate, and there is still a need to save electricity in the future. In doing so, it is important to consider how much of the electricity savings achieved so far will continue, in considering future energy-saving policies.

Table 3.1. Power-saving target for each region

	2011	2012	2013 -
Hokkaido	-	7%	-
Tohoku	15%	-	-
Kanto	15%	-	-
Hokuriku	-	-	-
Tokai	-	-	-
Kinki	10%	10%	-
Tyugoku	-	-	-
Shikoku	-	5%	-
Kyusyu	-	10%	-
Okinawa	-	-	-

Source: http://www.kantei.go.jp/jp/headline/summer2012_denryoku.html#c2_2.

So far, analysis on the continuity of energy-saving behavior in Japan has been insufficient. Nishio and Ofuji (2014), through a continuous questionnaire survey of households served by Tokyo Electric Power Company and Kansai Electric Power Company from 2011 to 2013 pointed out a decline in the implementation rate of measures, although the power-saving rate has been maintained. Furthermore, from the questionnaire survey of households served by the Tokyo Electric Power Company and the Kansai Electric Power Company, Tanaka and Ida (2013) pointed out that the power-saving behavior in 2012 would decrease as compared with the previous year. All of these researches point out the conservation consciousness and the secular change of the power-saving behavior by questionnaire; however, the verification of the power-saving results is not sufficient. Nishio and Ofuji (2014) used self-reported electricity usage data, and did not use the actual electricity usage amount. It is well known that self-reported data tend to overestimate the amount of energy reduction (

Joskow and Marron 1992). Furthermore, Tanaka and Ida (2013) have not verified continuity from the viewpoint of the energy-saving rate.

In addition, these studies cover only two areas in the Kanto Electric Power Company and the Kansai Electric Power Company, and other areas have not been verified.

This study utilized published data related to household electricity usage fee from the household survey conducted by Statistics Bureau of the Ministry of Internal Affairs and Communications for 10 regions in Japan (Hokkaido, Tohoku, Kanto, Hokuriku, Tokai, Kinki, China, Shikoku, Kyushu, Okinawa). We use these data to verify continuity of energy-saving behavior among Japanese households.

The structure of this paper is as follows. The next section present previous studies on continuity of household electricity-saving behavior in the international context. Section 3 discusses the data used in this study and the model used for the analysis. Sections 4 and 5 examine the estimation results and the continuity of the power-saving behavior based on these results, respectively. Section 6 concludes.

3.2. LITERATURE REVIEW

Reiss and White (2008) examined the effect of government's request to save electricity on a large scale. During the power crisis that occurred in the state of California in the United States from 2000 to 2001, they clarified the effect of energy saving of the household sector on the impact of the surge in electricity prices immediately after the crisis and the subsequent impact of power-saving demands. In California, the electricity price soared shortly after the summer of 2000 when the power crisis occurred. Thereafter, prices were restored due to price restrictions, and a large-scale energy-saving request was made by the government. In their research, 70,000 households were randomly extracted from San Diego, California, and electricity usage data for five years from 1997 (before the power crisis) to April 2002 (after the power crisis) were used to verify the response to price and the effect of power-saving request after pricing was implemented. As a result, the electricity-saving effect due to the sharp rise in electricity price was over 13% in 60 days, and 7% saving effect was confirmed for the subsequent power-saving request for six months. Their research is an effective study that showed public appeal for power saving has a tremendous effect even in the absence of monetary incentives. However, since this study considers a short period of six months after power-saving request, the continuity of the effect cannot be sufficiently verified due to restriction of observation. Numerous other studies have been conducted on the California power crisis. Ofuji and Nishio (2011) examined the effects of the public awareness on power saving in the household sector, while Ofuji and Kimura (2011) described the effect of the 20/20 program incentive system. Goldman et al. (2002) examined the household electricity-saving effect in California, where more than 200 electricity saving programs were implemented. As a result, in the summer (spring), 8.4% (4.8%) peak power and 6.1% (3.7%) electricity saving were observed. Furthermore, the power-saving effect shows that half of it has been sustained the following year.

Likewise, in the Alaska state of the United States, due to the avalanches that occurred in April 2008, a power crisis began with the collapse of the power transmission tower, and consequently, a large power-saving policy was implemented. Using a household survey, Leighty and Meier (2011) showed that 25% consumption reduction was achieved in 45 days immediately after the crisis, and in eight months after restoration, an effect of 8% on average remains. They pointed out that these are due to behavioral changes toward power saving and the effect of replacement to energy-saving equipment.

While these cases are continuous for about a few months to a year, Gerard (2013) observed the power-saving effect for 10 years after the 2001 power crisis in Brazil. In Brazil, hydropower generation was mainstream; thus, the drought that occurred in 2001 brought a power crisis. At this time, the government issued a crisis declaration, raised the electricity price, set the power-saving target, implemented a remuneration system for energy saving beyond the target value in the household sector, and increased the public awareness on power saving. Consequently, on average, about 25% demand control was achieved for nine months after the power crisis declaration in May 2001. In addition, after nine years, demand suppression of about 12% was observed. From this, he evaluated that about half of the initial power-saving effect remained. However, his research is an evaluation includes all sectors, not just the household sector, and it is unclear how the effect is brought about.

In Japan, there are also some studies that examined the effect of summer electricity-saving demands after the Great East Japan Earthquake in March 2011. Nishio and Ofuji (2012, 2013, 2014) conducted a questionnaire survey for families residing in areas served by the Kanto Electric Power Company and Kansai Electric Power Company, where the influence of power shortage was serious. From their survey data, they analyzed the energy conservation rate, consciousness to save electricity, and implementation rate of specific efforts to save electricity of the household sector. As a result of targeting more than 1,000 households in the follow-up survey for three years, in 2013, after the earthquake, the power-saving rate (compared to 2010) in summer (July–September) in areas served by Tokyo Electric Power Company and Kansai Electric Power Company remained about 10%. On the other hand, the implementation rate of energy-saving measures has declined in 2013, and it has been reduced to about 60 to 80% of the most active year.[10] In addition, the consciousness to save electricity is also declining. As for the motive for power saving, the normative motivation for solving the power shortage was high

---

[10] Particularly, the implementation rate of "use time reduction of air conditioner" decreased conspicuously. In the house inside TEPCO, its implementation rate has declined to about half of 2011 in 2013. Meanwhile, when the refrigerator is separated from the wall, after setting the brightness setting etc. of the television, it is also clear that these behavior are easy to continue without special consciousness.

immediately after the disaster; on the contrary, the incentive to reduce electricity price (economic motive) has been confirmed to rise. These findings are similar to those pointed out by Tanaka and Ida (2013). Furthermore, Nishio and Ofuji (2014) have verified the effect of renewal of household appliances, and a significant power-saving effect was confirmed in the replacement households. From these results, they point out that the continuous energy-saving effect is largely attributed to the switch to energy-saving equipment. The sample size of their research is large, and it is verified based on the follow-up survey for three years from 2011 to 2013. However, in response to the questionnaire, respondents self-reported electricity usage; therefore, there is concern that the reduction effect will be overestimated (Joskow and Marron 1992). In addition, Fujimi and Chang (2014) verified the power-saving effect for companies. To date, to the best of the author's knowledge, previous researches that verified the power-saving effect and its continuity used highly objective data published in Japan. In this study, we obtain home electricity usage monthly data (January 2006– February 2015) in 10 regions of Japan from the results by items of household survey of Ministry of Internal Affairs and Communications, considering household attributes and climate factors. We estimate the effect of power-saving request for each region and the continuity of the power-saving effect by regression analysis.

3.3. EMPIRICAL ANALYSIS

3.3.1. Data

In order to analyze the effect and continuity of the power-saving request issued in the regional unit, this study uses government data. First, from the "Family Income and Expenditure Survey" of the Ministry of Internal Affairs and Communications Statistics Bureau, we obtained the electricity data (monthly) tabulated by region from January

2003 to December 2016. Here, while electricity-saving requests are targeted for peak hours (daytime during weekdays), this study considers the monthly electricity bill. According to Nishio and Ofuji (2011), there are many households that are conscious of electricity conservation even at times other than peak hours (nighttime and holidays). Therefore, household electricity-saving behavior is expected to be implemented not only in the peak time zone but also in the non-peak time zone. Thus, it seems that there is certain significance in verifying the effect even on a usage basis. Furthermore, since this study uses data on royalties instead of usage, in order to adjust the price fluctuation, we calculated the electricity usage fee for each month based on the price of 2015 from "Consumer price index" of the Ministry of Internal Affairs and Communications Statistics Bureau.

Figure 3.1 shows the trend of real electricity usage charges in the summer (July–September) in 10 Japanese regions (Hokkaido, Tohoku, Kanto, Hokuriku, Tokai, Kinki, Tyugoku, Shikoku, Kyushu, Okinawa). In Japan, the electricity usage fee can be obtained on the meter reading day. From Figure 3.1, it can be seen that the electricity usage fee is approximately 9,000 yen per month; however, there are variations from year to year. Here, it is estimated that about 60% of the electricity usage fee in the summer is due to the use of air conditioners (Ministry of Economy, Trade and Industry, 2013), and the influence of the change in outdoor temperature will be great. Figure 3.2 shows the trend of the average outdoor temperature in summer season (difference from normal average) in the 10 areas. As you can see from Figures 3.1 and 3.2, when the average outdoor temperature is low in summer, electricity usage is also reduced; conversely if the average temperature is high, the electricity usage tends to increase. For example, in Figure 3.2, average temperatures in 2003, 2009, 2014, and 2015 are both below normal average, and the electricity usage fee for those years have decreased compared to the years before and after (see Figure 1). On the contrary, it can be seen from Figure 1 that the electricity usage fee is significantly increasing in the hot summer days like 2010. Figures 3.3, 3.5, 3.7, 3.9,

3.11, 3.13, 3.15, 3.17, 3.19, 3.21 shows the trends of electricity usage fee in summer in each area, and Figures 3.4, 3.6, 3.8, 3.10, 3.12, 3.14, 3.16, 3.18, 3.20, 3.22 shows the trends of the average outdoor temperature in summer (difference from normal average) in each area. From each figure, it can be confirmed that the use of electricity is decreasing in all areas except Okinawa since 2010, where the temperatures remained high. While this may be due to the effect of the power-saving request mentioned above, it seems the electricity usage follows a similar pattern as the average temperature. Therefore, electricity usage seems to be decreasing due to the power-saving request. However, the temperature is also decreasing in most areas. Therefore, there is a possibility that the effects are mixed. In order to isolate and analyze such complex factors, regression analysis is considered to be effective (Davis et al. 2014).

Another factor can be considered to explain power-saving behaviors in each region after 2011. The amount of thermal power generation was increased to compensate for the shortage of electricity supply due to shutdown of nuclear power plants, which caused a rise in fuel cost, thereby pushing the price of electricity upward. Figure 3.23 shows the transition of the electricity price index for households in 10 regional power companies in Japan. From this figure, before the 2011 East Japan Great Earthquake occurred, the electricity price index in the summer was relatively stable. However, it can be seen that the electricity price index is increasing rapidly since 2012. In particular, households in the Kanto Electric Power Company and the Kansai Electric Power Company have confirmed a surge in electricity price of 30% to 35% compared with 2010. The response of electricity demand to electricity price can be grasped to some extent using price elasticity. Okajima and Okajima (2013) showed that the electricity price elasticity of Japanese households is - 0.397 in the short term and - 0.487 in the long term, indicating that it is inelastic but not negligibly small. Therefore, the sharp rise of the price after 2011 is considered to be a factor that suppresses the electricity demand of households slightly.

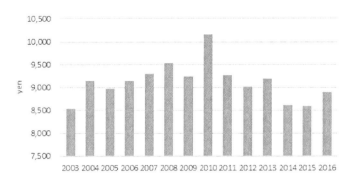

Figure 3.1. The change of household electricity bills in summer (July, August, September).

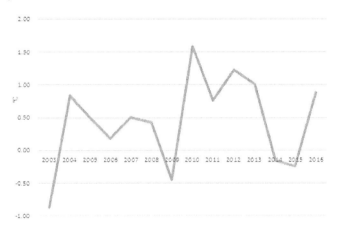

Figure 3.2. Average outside air temperature in summer (difference from normal average).

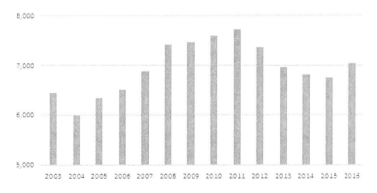

Figure 3.3. Household electricity bill in summer (Hokkaido).

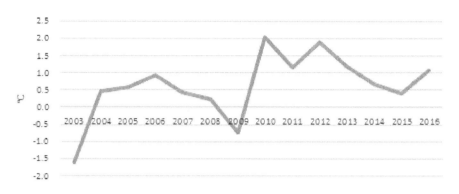

Figure 3.4. Average outside air temperature in summer (difference from normal average) Hokkaido.

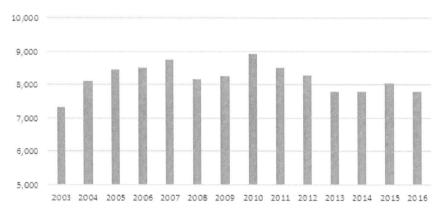

Figure 3.5. Household electricity bill in summer (Tohoku).

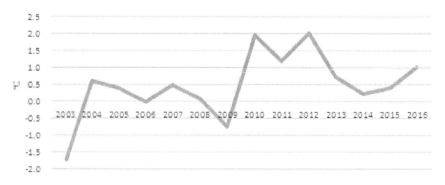

Figure 3.6. Average outside air temperature in summer (difference from normal average) Tohoku.

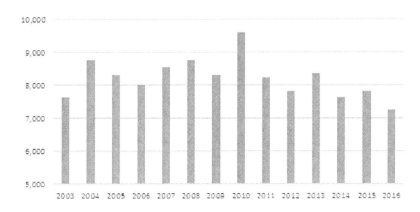

Figure 3.7. Household electricity bill in summer (Kanto).

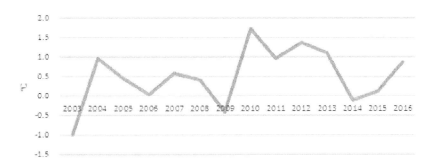

Figure 3.8. Average outside air temperature in summer (difference from normal average) Kanto.

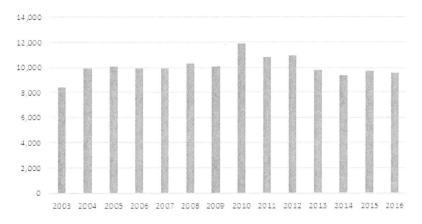

Figure 3.9. Household electricity bill in summer (Hokuriku).

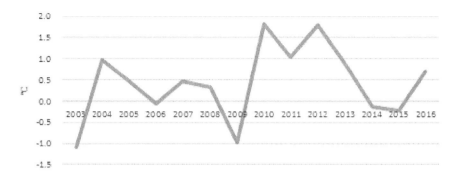

Figure 3.10. Average outside air temperature in summer (difference from normal average) Hokuriku.

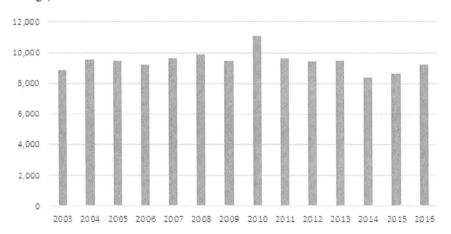

Figure 3.11. Household electricity bill in summer (Tokai).

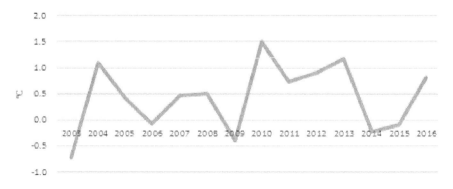

Figure 3.12. Average outside air temperature in summer (difference from normal average) Tokai.

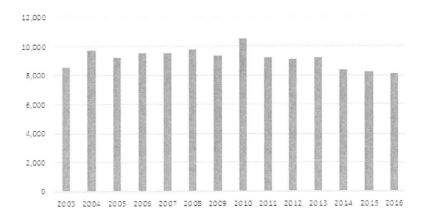

Figure 3.13. Household electricity bill in summer (Kinki).

Figure 3.14. Average outside air temperature in summer (difference from normal average) Kinki.

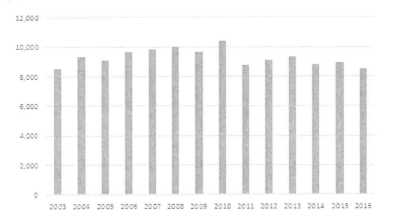

Figure 3.15. Household electricity bill in summer (Chugoku).

Figure 3.16. Average outside air temperature in summer (difference from normal average) Chugoku.

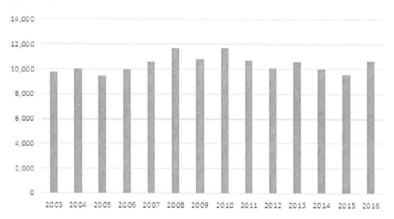

Figure 3.17. Household electricity bill in summer (Shikoku).

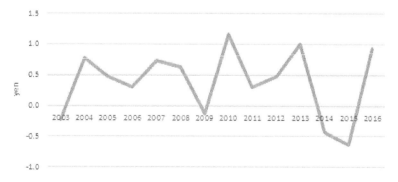

Figure 3.18. Average outside air temperature in summer (difference from normal average) Shikoku.

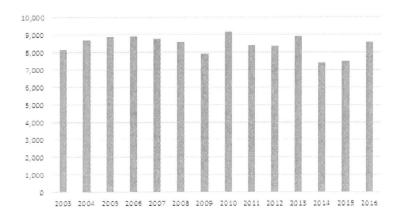

Figure 3.19. Household electricity bill in summer (Kyusyu).

Figure 3.20. Average outside air temperature in summer (difference from normal average) Kyusyu.

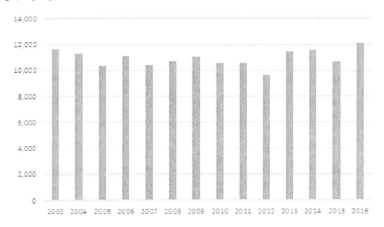

Figure 3.21. Household electricity bill in summer (Okinawa).

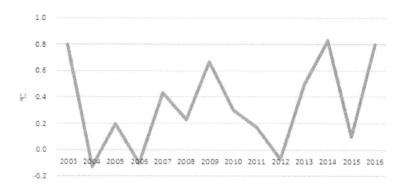

Figure 3.22. Average outside air temperature in summer (difference from normal average) Okinawa.

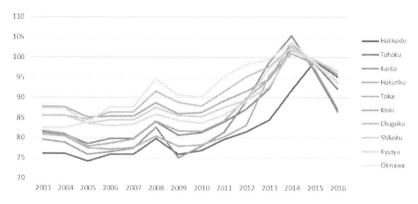

Figure 3.23. Trend of Electricity Price Index (100 in 2015).

As shown in Table 3.1, power-saving requests are set based on the amount of electricity used in 2010. To compare with the power-saving request level in Table 3.1, Figure 3.24 shows the trend of the power-saving rate in each year based on the summer of 2010 except Okinawa (Because Okinawa has no nuclear power, the problem of power shortage does not occur). It can be seen that the power-saving rate is widely distributed from 6.1% to 19.1%. As 2010 was a hot summer, it seems that the power-saving rate was negative in 2010 and positive in all other years. It is expected that the power-saving rate depends not only on request for energy saving but also on changes in temperature. Figure 3.25 shows the trend of the average outdoor temperature (difference from 2010) in summer season. Corresponding to Figure 3.24, it can be seen that in low-temperature cities, the power-saving rate rises and in high cities, the power-saving rate decreases. However, when comparing the temperature from 2005 to 2009 with that from 2011 to 2013, the energy conservation ratio is almost the same despite the fact that in the latter period, the temperature was about 1°C higher. Therefore, after the earthquake, the energy conservation results can be seen to be positive even considering the temperature. However, as shown in Figure 3.23, this result may be attributed the factor of electricity price rise, which is included in this result.

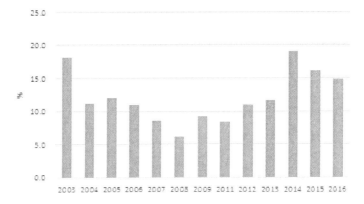

Figure 3.24. Trend of energy saving rate (based on 2010).

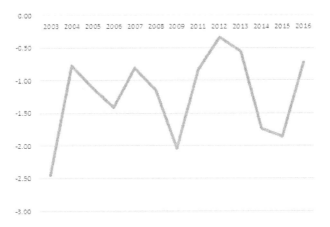

Figure 3.25. Trend of temperature in summer (difference from 2010).

From the above, it is clear that after the earthquake in 2011, the power-saving rate is improved. However, we cannot judge whether or not the effect of power-saving request is effective and its magnitude only by data, as the influence of temperature and price rise is included in addition to the power-saving request made by the government. Therefore, in order to verify the effect of the power-saving request, it is necessary to analyze by regression analysis that controls temperature, price, and other factors.

3.3.2. Regression Model

As described in the previous section, we need to verify the reduction of electricity use in the household sector after 2011, in consideration of factors other than the request for power saving. We will use the regression model to verify mainly the effects of 1) power-saving request; 2) outdoor temperature change; and 3) the effect of increase in electricity price. The regression model used in this study is as follows:

$$REDUCTION_{i,t} = \alpha + \beta_1 P\_ELEC_{i,t} + \beta_2 TEMP_{i,t} + \beta_3 REQUEST_{i,t} + \gamma X_{i,t} + \eta T_i + \varepsilon_{i,t},$$

$$(3.1)$$

where REDUCTION is the power-saving rate, P_ELEC is the electricity price, TEMP is the outdoor temperature, and REQUEST is the electricity-saving request. Here, i represents the area (Hokkaido, Tohoku, Kanto, Hokuriku, Tokai, Kinki, Chugoku, Shikoku, Kyushu, Okinawa), and t represents time (January 2003–December 2016). Since the targets in the power-saving request were based on electricity usage data for 2010, the power-saving rate in area i in the year t is calculated based on the following formula (The transition of the power-saving rate in Figure 4 also uses this equation).

$$REDUCTION_{i,t} = -\frac{(ELEC_{i,t}-ELEC_{i,2010})}{ELEC_{i,2010}}, \qquad (3.2)$$

where ELEC represents the electricity usage fee for the year t in i region. The power-saving rate given by Eq. (3.2) is positive if households conserve electricity on the basis of 2010, and becomes negative if it increases electricity. Electricity-saving request have been implemented in each region except Okinawa since the summer of 2011. In order to verify the electricity-saving request effect, a dummy was set for each area (with or without target). Furthermore, because the average temperature matches the power-saving rate, the difference between the temperature of each year and the temperature of 2010 was taken. The electricity price is calculated from the consumer price index. X includes the total expenditure (logarithmic value), the number of households, the number of people aged 65 years and over, and the homeownership ratio as other variables. T is a time trend, and it can capture the secular change that cannot be explained with the explanatory variable above. ε represents an error term.

3.4. ESTIMATION RESULTS

Since the model of this study uses panel data, three models—pool estimation model, fixed effect model, and fluctuation effect model— were considered. From the results of F test, Hausman test, Breusch and Pagan test, fixed effect>fluctuating effect>pool order; thus, the results of the fixed effect model was adopted in this study. Table 3.2 shows estimation results for the fixed effect model. The dependent variable is the power-saving rate calculated based on Eq. (3.2). Models (1) and (2) include estimates of the effect of power-saving request in the summer after the Great East Japan Earthquake without considering the changes in temperature and electricity price. Models (3) and 4) are the results obtained on considering changes in temperature and electricity price. As a summer dummy after the earthquake, the summer (July to September) after 2011 is set to 1 and the others are set to 0 as the explanatory variable in models (1) and (3). Furthermore, models (2) and (4) were estimated by adding the dummy variables every summer of each year, and the change of influence every summer was analyzed.

The results of model (1) show that all the variables are statistically significant. Here, when the sign of the estimation coefficient is positive, it indicates that power saving is achieved, whereas if it is negative, electricity usage is increased. As the estimation coefficients of income level, homeownership ratio, and size of family are negative and significant, this implies that as the income level increases, the higher the ownership ratio and the higher the size of family, the more the use of electricity. As the number of family members increases, the number of people who need to cooperate in saving electricity increases, and thus, it may be difficult to achieve power saving. Furthermore, since the coefficients of the time trend are negative and significant, it indicates that the power-saving rate has gradually declined from 2003 to 2016.

Next, we examine the effect of power-saving request. The coefficient of summer dummy after the earthquake was positive and significant in model (1). This seems to indicate that the household sector also helped save electricity as a result of the power-saving request implemented by the government since the Great East Japan Earthquake, indicating that the power-saving level has risen. In model (2), when estimating it separately for summer dummies of each year, after 2012, the coefficients were all positive and significant.

Table 3.2. Estimation results

		Model (1)		Model (2)		Model (3)		Model (4)	
price of electricity	logarithm	-		-		99.431	***	103.202	***
						[9.122]		[9.834]	
temperature difference from 2010 (summer)	°C	-		-		-1.955	***	-2.072	***
						[0.265]		[0.308]	
temperature difference from 2010 (winter)	°C	-		-		2.349	***	2.366	***
						[0.293]		[0.293]	
summer dummy after earthquake disaster		6.133	***	-		3.646	***	-	
		[0.703]				[0.689]			
summer dummy 2004		-		0.497		-		-2.210	
				[1.545]				[1.452]	
summer dummy 2005		-		2.361		-		1.116	
				[1.538]				[1.459]	
summer dummy 2006		-		0.970		-		-0.848	
				[1.529]				[1.473]	
summer dummy 2007		-		0.206		-		0.335	
				[1.522]				[1.440]	
summer dummy 2008		-		-1.336		-		-3.528	**
				[1.519]				[1.445]	
summer dummy 2009		-		1.687		-		0.148	
				[1.519]				[1.536]	
summer dummy 2011		-		0.798		-		0.548	
				[1.521]				[1.443]	

Table 3.2. (Continued)

		Model (1)		Model (2)		Model (3)		Model (4)	
summer dummy 2012		-		5.429	***	-		4.704	***
				[1.523]				[1.423]	
summer dummy 2013		-		4.864	***	-		1.859	
				[1.527]				[1.437]	
summer dummy 2014		-		10.702	***	-		3.376	**
				[1.546]				[1.552]	
summer dummy 2015		-		9.306	***	-		3.910	**
				[1.563]				[1.554]	
summer dummy 2016		-		7.043	***	-		7.156	***
				[1.577]				[1.494]	
income	logarithm	-37.695	***	-37.018	***	-33.741	***	-34.692	***
		[5.861]		[5.880]		[5.465]		[5.484]	
homeownership ratio	%	-0.238	***	-0.231	***	-0.213	***	-0.215	***
		[0.071]		[0.071]		[0.066]		[0.066]	
size of family		-25.246	***	-23.500	***	-32.796	***	-30.960	***
		[8.292]		[8.313]		[7.709]		[7.756]	
number of people aged 65 years and over		13.962	***	11.577	***	7.274	**	5.608	
		[3.945]		[4.004]		[3.690]		[3.754]	
time trend		-1.080	***	-1.029	***	-1.742	***	-1.744	***
		[0.183]		[0.185]		[0.183]		[0.187]	
constant		307.270	***	298.707	***	123.778	***	117.222	***
		[41.590]		[41.743]		[41.278]		[42.065]	
R square	overall	0.001		0.004		0.035		0.039	
	within	0.112		0.130		0.237		0.248	
	between	0.277		0.302		0.206		0.237	
number of observations	whole	1,560		1,560		1,560		1,560	
	region	10		10		10		10	

"***," "**," and "*" indicate 1%, 5%, and 10% significance level, respectively.

As the coefficients before the earthquake are not significant, we can see that there was a certain effect of the government's request for energy saving. However, as we have verified in Section 3.3.1, there has been a drastic rise in electricity prices after 2011 and a decline in the average temperature in the summer season; consequently, it is difficult to judge that all of the improvement in the power-saving rate after 2011 is attributed to the effect of the power-saving request. Therefore, in order to eliminate the rise in the electricity price and the influence of the temperature, we include these in the estimation model as explanatory variables and conducted the estimation again. The results are shown in models (3) and (4). The signs and significance results of the variables included in models (3) and (4), such as total expenditure, homeownership ratio, and size of family, are almost the same as in models (1) and (2).

Next, we look at the impact of the rise in electricity prices. The power price estimation coefficients of models (3) and (4) are positive and statistically significant. As the price of electricity rises, it is expected that households will save electricity consumption and save energy. As shown in Figure 3.3, the increase in the share of thermal power generation after the Great East Japan Earthquake has caused the procurement cost of fossil fuel to rise, which has gradually been added to the electricity price. Previous studies have shown that price elasticity of household electricity is low. However, the electricity price increase is often reported largely in the media, and recently, the fixed price purchase system of renewable energy has also become popular, and the burden of purchase price imposed on general households increases year by year. Therefore, in recent years, the price elasticity of household electricity demand may be larger. On the other hand, the variable of the temperature difference is negative in the summer and positive in the winter, both of which are statistically significant. Since this temperature difference is the difference in temperature from 2010, it was shown that there is a tendency for electricity usage to increase in summer as power saving becomes more difficult when the temperature falls compared with 2010. On the other hand, in the winter, if the temperature rises above the 2010 temperature, heating energy will be suppressed and it will be easier to save electricity. It can be said that the sign of the estimated parameter of the temperature is a reasonable result.

Finally, we examine the extent to which the power-saving request was effective after controlling the rise in electricity prices and the influence of the temperature. First, in model (3), the coefficient of the summer dummy after the earthquake is positive and statistically significant, which is the same as the result of model (1). In other words, it can be concluded that even in a situation where the increase in electricity price and the change in temperature are controlled, the electricity-saving request implemented by the government brings about the power-saving promotion effect in the household sector. On the other hand, the magnitude of the estimation coefficient is 6.133 in model (1) and 3.646 in model (3). Thus, the influence becomes slightly smaller when the influence of price and temperature is eliminated. From this result, it can be said that there is a possibility of excessively evaluating the effect of the power-saving request if judged from only the data of the power-saving rate without considering the price rise and change of the temperature.

Next, we will examine the transition of the power-saving request effect in the summer of each year after the earthquake from the result of model (4) with the summer dummy of each year. Except for 2013, the estimated parameters are all positive and statistically significant after the disaster. In particular, it can be confirmed that there was a large power-saving effect in 2012 when the demand for power saving began in full scale nationwide. Furthermore, even after 2013, when targets for power-saving requests were not set, the power-saving effect continues. In particular, the power-saving rate after 2014 has a level that is not significantly different from the 2012 power-saving rate. This is consistent with the result of Nishio and Ofuji (2014). Their questionnaire survey also shows that the energy-saving rate in summer is almost maintained from 2011 to 2013. On the other hand, as mentioned in Section 3.2, in the years after the earthquake, it is possible that the switch to energy-saving home appliances progressed, which may have supported the lowering of the power-saving rate due to the reduction of the power-saving behavior. In this study, since the replacement behavior of households using energy-saving home appliances is not grasped, it is fully conceivable that the effect of such replacement is still included.

3.5. CONTINUITY OF POWER-SAVING EFFECT BY REGION

As shown in Table 3.1, the presence or absence of power-saving request and its target level are different according to year and region. In this section, we will examine the presence or absence of power-saving effect by region and its continuity. In the model used for analysis, the power-saving request dummy is divided into regions, and the effect in each region and the transition of the power-saving effect every year are estimated. Table 3.3 presents the estimation results. In the model selection, since the fixed effect model was selected, only the result is posted.

Models (5) and (7) have summer dummies in nine areas, and models (6) and (8) are the result of dividing summer dummy into dummies of year by year. The results of model (5) were almost the same as those shown in Table 3.2, including the total expenditure, homeownership ratio, size of family, number of people aged 65 years and over, and time trends. As a result of the power-saving request, the summer dummy after the earthquake in each area has become significant in all areas except Hokkaido, and it can be confirmed that there was a power-saving effect. Moreover, from the results of model (6), significant power-saving effects can be observed for many years. Meanwhile, there are variations in the level of power-saving effect for each region and year. For example, in 2012, we confirmed a large difference between -0.19% (not significant) of Hokkaido and 11.04% in Kanto, and in 2013, from 0.47% (not significant) in Kyushu to 8.47% in Kinki. These may depend on power-saving demands and the level of targets.

Models (7) and (8) are the result of controlling the increase in electricity price and change in temperature. The estimation coefficients of electricity price and temperature (summer and winter) are significant, indicating that there was an influence on the power-saving rate. Moreover, compared with models (5) and (6), the size of the summer dummy by region is small, so if we do not consider these factors, there is a possibility to overestimate the power-saving effect.

Here, using the result of model (8), we look at the effect of numerical targets of power-saving request. From Table 1, in the summer of 2011, numerical targets of 15% in the Tohoku and Kanto regions and 10% in the Kinki district were set. In Table 3.3, regional dummies for summer 2011 were -0.77%, 5.55%, and 5.14%, respectively, which were below the numerical targets. In Section 3.3, we mentioned that the object of analysis in this study is not the peak power base, but the amount of electricity in all time zones. Since the numerical targets of the power-saving request are set for the peak electric power, it is not uncommon that there will be a difference from the electricity charges including other time zones.

Table 3.3. Estimation results (by prefecture)

		Model (5)	Model (6)	Model (7)		Model (8)	
price of electricity	logarithms	-	-	98.833	***	101.287	***
				[9.157]		[9.774]	
temperature difference from 2010	°C	-	-	-2.018	***	-1.714	***
(summer)				[0.265]		[0.277]	

		Model (5)		Model (6)		Model (7)		Model (8)	
temperature difference from 2010 (winter)	°C	-		-		2.363	***	2.346	***
						[0.293]		[0.291]	
Hokkaido dummy (summer)		2.070		-		-0.624		-	
		[2.086]				[1.944]			
2011				-7.275				-6.864	
				[4.801]				[4.493]	
2012				-0.189				0.924	
				[4.743]				[4.436]	
2013				5.894				4.959	
				[4.759]				[4.452]	
2014				7.440				2.963	
				[4.777]				[4.482]	
2015				5.588				-1.803	
				[4.804]				[4.532]	
2016				3.809				-0.590	
				[4.761]				[4.463]	
Tohoku (summer)		4.977	**	-		2.025		-	
		[2.114]				[1.972]			
2011				-1.563				-0.768	
				[4.750]				[4.445]	
2012				3.756				4.062	
				[4.762]				[4.452]	
2013				8.432	*			5.295	*
				[4.755]				[4.455]	
2014				9.635	**			1.080	
				[4.786]				[4.528]	
2015				5.242				-0.129	
				[4.760]				[4.475]	
2016				6.784				5.905	
				[4.811]				[4.503]	
Kanto (summer)		10.236	***	-		6.779	***	-	
		[2.092]				[1.955]			
2011				4.818				5.545	
				[4.752]				[4.450]	
2012				11.036	**			9.394	**
				[4.743]				[4.434]	

Table 3.3. (Continued)

	Model (5)		Model (6)		Model (7)		Model (8)	
2013			7.444				1.569	
			[4.756]				[4.469]	
2014			15.218	***			5.226	
			[4.753]				[4.517]	
2015			9.757	**			4.679	
			[4.771]				[4.485]	
2016			16.229	***			17.832	***
			[4.780]				[4.483]	
Hokuriku(summer)	4.097	*	-		1.955		-	
	[2.114]				[1.975]			
2011			-1.297				-1.403	
			[4.743]				[4.437]	
2012			-5.168				-4.128	
			[4.775]				[4.465]	
2013			6.565				4.280	
			[4.777]				[4.470]	
2014			11.054	**			4.978	
			[4.780]				[4.506]	
2015			7.407				3.607	
			[4.788]				[4.510]	
2016			8.909	*			8.424	*
			[4.786]				[4.485]	
Tokai (summer)	9.755	***	-		6.360	***	-	
	[2.487]				[2.317]			
2011			3.928				4.477	
			[4.745]				[4.442]	
2012			8.222	*			6.260	
			[4.746]				[4.438]	
2013			10.192				7.215	
			[4.750]				[4.445]	
2014			20.055	**			11.022	**
			[4.763]				[4.514]	
2015			15.624	***			10.541	***
			[4.759]				[4.475]	
2016			10.380	**			12.343	***
			[4.784]				[4.490]	

	Model (5)		Model (6)		Model (7)		Model (8)	
Kinki (summer)	8.382	***	-		5.992	***	-	
	[2.488]				[2.312]			
2011			4.032				5.141	
			[4.748]				[4.448]	
2012			7.767				9.183	**
			[4.750]				[4.445]	
2013			8.475	*			4.177	
			[4.755]				[4.457]	
2014			17.707	***			9.600	**
			[4.752]				[4.492]	
2015			17.198	***			10.304	**
			[4.758]				[4.491]	
2016			19.094	***			17.704	***
			[4.769]				[4.461]	
Tyugoku (summer)	6.439	***	-		3.818	**	-	
	[2.088]				[1.960]			
2011			4.718				3.364	
			[4.777]				[4.472]	
2012			4.866				3.477	
			[4.749]				[4.440]	
2013			4.180				1.343	
			[4.769]				[4.463]	
2014			8.866	*			2.768	
			[4.752]				[4.488]	
2015			7.790				3.959	
			[4.765]				[4.505]	
2016			10.145	**			11.651	***
			[4.767]				[4.476]	
Shikoku (summer)	4.008	*	-		1.626		-	
	[2.083]				[1.944]			
2011			-0.441				-1.170	
			[4.760]				[4.457]	
2012			5.732				5.297	*
			[4.752]				[4.446]	
2013			0.570				-0.193	
			[4.750]				[4.440]	
2014			6.938				0.715	
			[4.753]				[4.476]	

Table 3.3. (Continued)

		Model (5)		Model (6)		Model (7)		Model (8)	
2015				10.868	**			6.007	
				[4.753]				[4.475]	
2016				2.208				2.153	
				[4.759]				[4.450]	
Kyusyu (summer)		7.097	***	-		5.017	***	-	
		[2.082]				[1.942]			
2011				1.403				2.277	
				[4.748]				[4.447]	
2012				5.593				4.956	
				[4.751]				[4.445]	
2013				0.474				-1.689	
				[4.744]				[4.439]	
2014				17.563	***			11.299	**
				[4.754]				[4.476]	
2015				15.607	***			10.794	**
				[4.763]				[4.495]	
2016				4.408				5.817	
				[4.761]				[4.457]	
income	logarithms	-37.400	***	-35.950	***	-33.627	***	-32.415	***
		[5.893]		[5.874]		[5.486]		[5.515]	
homeownership ratio	%	-0.214	***	-0.235	***	-0.201	***	-0.222	***
		[0.072]		[0.074]		[0.066]		[0.069]	
house size		-26.666	***	-26.244	***	-34.558	***	-33.516	***
		[8.478]		[8.549]		[7.872]		[8.007]	
number of people over 65 years old		13.921	***	11.214	***	7.277	**	5.796	
		[3.971]		[4.076]		[3.708]		[3.838]	
time trend		-1.078	***	-1.101	***	-1.748	***	-1.783	***
		[0.186]		[0.187]		[0.185]		[0.188]	
constant		308.387	***	302.635	***	128.988	***	117.210	***
		[42.367]		[42.551]		[41.891]		[42.905]	
R square	overall	0.003		0.015		0.040		0.053	
	within	0.114		0.166		0.241		0.273	
	between	0.203		0.215		0.164		0.182	
number of observations	whole	1,560		1,560		1,560		1,560	
	region	10		10		10		10	

"***," "**," and "*" indicate 1%, 5%, and 10% significance level, respectively.

Next, I will look at the result of 2012. The set energy saving numerical targets were 7% in Hokkaido area, 10% in Kinki area, 5% in Shikoku area and 10% in Kyushu area. The results of the estimation coefficients in Table 3.3 are 0.92%, 9.18%, 5.30%, and 4.96%, respectively. From this, it can be seen that Shikoku region alone shows a power-saving effect exceeding numerical targets. In this year, since power-saving requests (including areas not accompanied by numerical targets) were made nationwide, there is a possibility of saving electricity in other areas as well. For example, 9.39% in the Kanto region and 6.26% in the Tokai region were conserved. From this, although the industrial sector and the public sector have primarily played a major role in energy conservation, the consciousness of energy conservation and actions may have appeared in the households as well.

In the summer of 2013, no targets were set, and calls for saving energy in a reasonable range were invited. As a result, unlike the summer of 2012, the power-saving rate of the previous year is not achieved in each region, and only in the Tohoku region, a statistically significant power-saving rate is shown. Even in the summer dummy in 2013 in Table 3.2, the figures have dropped to less than half compared to the previous year (2012: 4.70%, 2013: 1.86%). However, although the power-saving rate has become smaller in many areas, it may be possible to view that part of the effect is continuing as the coefficient is positive.

In the summer of 2014, it was the first summer to come with zero operation of nuclear power, since Oishi nuclear power plant of Kansai Electric Power Company was suspended for inspection. However, the government implemented "Power Saving Request without Numerical Targets" and set no energy-saving targets. However, the electricity supply shortage was more severe than in 2013 due to zero nuclear power. Therefore, there is a high possibility that power-saving measures were taken in each area and each department from the previous year, which is also seen in the household sector. As shown in Table 3.3, a large power-saving effect was seen in five areas: 5.23% in the Kanto area, 4.98% in the Hokuriku district, 11.02% in the Tokai district, 9.60% in the Kinki district, and 11.30% in the Kyushu district. This level is nearly the same as in 2012, when consciousness of energy saving increased nationwide. Therefore, even without accompanying numerical targets, this high power-saving effect was obtained; so, it may be said that there is constant continuity of saving electricity. It is also possible that people's energy-saving behaviors have shifted toward more efficient ones since the earthquake. Frequent turning-off and standby power zero, among others, are slightly smaller compared to air conditioner usage time reduction, set temperature adjustment, replacement for energy-saving home appliances. However, with the request for energy saving every year, in the process of actually working on energy conservation, it may be possible to notice more efficient power saving (or to provide information by media, etc.), and the power-saving effect may have increased.

Past power saving numerical targets and nationwide demand for energy savings have encouraged people to conserve electricity, and it has become clear that the effect of that has continued for a long time through learning. This result shows that the environmental policy of saving energy is effective both in short term and medium to long term.

CONCLUSION

In this study, we examined the effectiveness of the power-saving request implemented by the government and its continuity in the nationwide power shortage problem owing to the nuclear power plant shutdown caused by the Great East Japan Earthquake that occurred in March 2011. Because power-saving requests include numerical targets and differences in target levels depending on region and year, it is necessary to verify the effect for each region and year. Therefore, panel data including the period from 2011 to 2016 after the earthquake and nine areas where power shortage actually occurred was obtained from the Ministry of Internal Affairs and Communications survey and analyzed. In addition, with electricity price increase since 2011 and temperature changes in the summer are included in changes in electricity usage, we will verify the effectiveness and continuity of the power-saving request by controlling these factors.

Based on the results in Sections 3.4 and 3.5, statistically significant power-saving effect was confirmed in the summer electricity-saving request after 2011 (even in the situation of controlling the increase in electricity fee and change in temperature). In addition, it was shown that possibly overestimating the effect of power-saving request without considering fee rise and temperature change. We also verified the power-saving effect by region and year. As a result, it turned out that the power-saving effect was dispersed for each area and year. Meanwhile, since the power-saving effect was maintained in some areas even after 2013 when the power-saving request ceased, we could also confirm the continuity of power saving.

Finally, we present two points on future tasks. The first point is the change in ownership type of electric appliances of each household. The holding situation of household appliances should have changed between 2003 and 2016, which is the analysis period of this paper, and if energy-saving home appliances become popular, it will lead to direct energy saving. Therefore, the analysis including the possession status and update information of home appliances will become necessary in the future. In Chapter 6 of this book, we verify the effect of saving electricity through replacement. The second point is the verification of the long-term effect. The tendency to shift to renewable energy by reducing nuclear dependence is a global trend. However, it takes a long time to spread, and during this time, it is necessary to rely on thermal power generation. The financial burden on companies and families is expected to increase in the future as the sources of fuel cost surge and renewable energy dissemination are gradually added to the electricity price. Therefore, the motive for power saving is expected to shift from the normative motivation of avoiding power shortage, as in the past, to the economic motivation of electricity saving. It is necessary to analyze the continuity of household electricity-saving behaviors in this movement.

Chapter 4

FIELD EXPERIMENT 1:
THE EFFECT OF A STEPWISE REWARD:
THE EFFECT OF ECONOMIC INCENTIVE
TO CONSERVE ELECTRICITY

ABSTRACT

In this chapter, we are examining whether it is possible to encourage energy saving behaviors at home by giving incentives. We conducted field experiments for 54 households in Matsuyama city, Ehime Prefecture, Japan for three months from November 2010 to January 2011. Participants set three levels of incentives (i) 0% - 9%, ii) 10% -19%, iii) 20% or more) according to the 3 month conservation rate. As a result, about 34% of households saved electricity, but households that saved 20% or more were as low as 3.8%. In addition, from the results of the preliminary questionnaire and post-questionnaires, it became clear that many households underestimate the difficulty of saving electricity. Furthermore, in the post-questionnaires, 33% of households responded that they do not take energy-saving behavior even if they give twice the incentive of this research (this chapter is a revision of Mizobuchi and Takeushi, 2012).

Keywords: electricity saving behavior, economic incentive, field experiment, household

4.1. INTRODUCTION

The aims of this chapter is to evaluate the effectiveness of introducing a stepwise economic incentive to encourage electricity conservation behavior, with the help of a field experiment that was conducted from November 2010 to January 2011, targeting Japanese households. This study is different from previous studies with respect to its method for recruiting participants of the experiment. As part of this study, we administered two questionnaire surveys, before and after the experiment, and used the responses from the subjects to examine the characteristics that contribute to the successful reduction in energy use. It must be noted that this study was conducted before the Great East Japan Earthquake in March 2011. Hence, the electricity-saving behavior of households was not influenced by massive energy conservation measures introduced in Japan after the nuclear disaster and the subsequent shutting down of nuclear power plants.

The study is organized as follows: Section 4.2 describes our field experiment, which uses a stepwise economic incentive for electricity conservation behavior. Section 4.3 shows the results of our empirical analysis. Section 4.4 discusses the marginal costs of electricity conservation behavior, and Section 4.5 describes some limitations. Section 4.6 presents our conclusion.

4.2. FIELD EXPERIMENT

4.2.1. Participated Households

There are about 482,000 people and about 221,000 households in Matsuyama city, Ehime prefecture in 2017. We selected seven business establishments in Matsuyama city and requested them to allow their employees to participate in our field experiment in November 2010. If a business establishment agreed to participate, all the employees of the firm were enrolled in the study. Fitty-three households were participated in our filed experiment. As a result, households were included in the experiment regardless of their environmental consciousness or motivation for energy conservation. Table 4.1 compares the socio-economic characteristics of our participated households with the other households in Matsuyama city and across the country. Except for gender rate and age, the characteristics of the sample households were similar to that of the general population. The average age of the sampled households (41.2 years) was lower than that of the general population. This is probably because the age variable for Matsuyama city and the rest of the country denotes the age of householder while in the case of the sampled households, it denotes the age of family member currently working in the participating firm. The non-inclusion of retired individuals in our sample could have also contributed to lowering the average age.

Table 4.1. The validity of our participated households

	Participation group	Matsuyama city	Japan
electricity consumption (yen)	10,954	9,159	9,423
- the average of Nov., Dec., and Jan. -			
gender rate (men/women)	57	47	49
age	41.2	56.3	55.9
family size	3.02	2.88	3.1
homeownership rate (%)	71.1	79.3	80.1
unmarried	32.1	-	27.1

Sources: Family Income and Expenditure Survey, Bureau of Statistics Japan (2010), A National Census of Japan (2010).

4.2.2. The Stepwise Economic Incentive and Experimental Procedure

Our field experiment was carried out for three months, from November 2010 to January 2011. Each enrolled household was entitled to receive a reward for reducing their electricity consumption to a level below the previous year's consumption for the same period. Electricity consumption was measured in kWh to ensure that the dependent variable, percentage change in total consumption, was comparable across households regardless of the energy fuels used. The power saving rate was calculated as follows:

$$Power\ saving\ rate\ (\%) = - \frac{(C_p - C_x)}{C_p} \times 100, \tag{4.1}$$

where C_x and C_p represent the electricity consumption during the experiment and during the same period in the previous year, respectively. If power saving rate was larger than 20% (*power saving rate* ≥ 20), households were given a reward (economic incentive) of 7,000 yen (about \$70). If the power saving rate was between 10% and 20% ($10 \leq power\ saving\ rate < 20$), households received 2,500 yen (about \$25), and if the power saving rate was between 0% and 10% ($0 < power\ saving\ rate < 10$), households received 500 yen (about \$5)[11]. Figure 4.1 describes the mechanism of this economic incentive.

For calculating power saving rate based on equation (4.1), we need the electricity consumption data of each participated household. Shikoku Electric Power Co. was the electricity supplier for all participating households. Participants could access their historical data on electricity consumption by logging onto the Shikoku Electric Power website, using their unique identification number. We collected these consumption data for the previous year from the households (printed web page) and used them as the measurement criteria. Additionally, for tracking ongoing consumption, every household was asked to send a copy of the monthly electricity consumption record sent to them by Shikoku Electric Power Co.

---

[11] We have assumed the following exchange rate: \$1 = 100 yen.

We did not divide the households into subgroups, on the basis whether to give incentives or not, to examine the effectiveness of the economic incentive. This is because a sample size of less than 20 tends to hamper the statistical validity of the results. Instead, we performed a regression analysis, with *power saving rate* as the dependent variable, to examine the influence of an economic incentive on the electricity conservation behavior.

Before and after the field experiment, we conducted questionnaire survey. In the pre-experiment questionnaire, we asked participants the following questions: their attitude to environmental issues, environmental awareness, attitude to electricity conservation, intended target level of electricity saving and the confidence to meet this target, types of electrical appliances in the household, and social, economic, and demographic details. In the post-experiment questionnaire, we asked participants about changes in their attitudes toward electricity conservation, the challenges in conserving electricity, the appropriateness of economic incentives, etc.

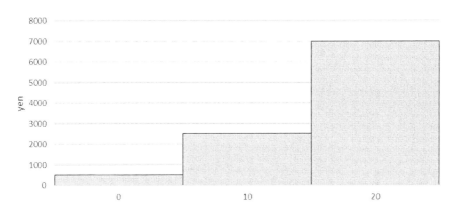

Figure 4.1. The mechanism of economic incentive of our field experiment.

The participated households were extracted as follows. We requested seven business firms, employing a total of 500 employees, to participate in our field experiment. These firms were selected because they satisfied certain conditions, e.g., the employees belonged to wide range of age groups, represented different types of family structures, and had different ownership patterns. We did not divulge any details about the field experiment while asking the business firms to participate in the study. Two of these firms agreed to participate, and a total of 55 employees or households were enlisted. However, employees who had moved house within a year were excluded from the study owing to the potential difficulties introduced by a change in the size of the house and the challenges in verifying electricity consumption data for the previous year. Thus, a total of 53 households were selected as the sample.

As the purpose of this study is to exclusively measure the effectiveness of economic incentives, other interventions for promoting electricity saving, such as feedback, were not used. Moreover, we did not request the participants to actively engage in electricity conservation. The study, thus, essentially examines changes in household energy-saving behavior in response to economic incentives.

4.2.3. Results of Our Field Experiment

After the end of our field experiment for 3 months, we got electricity consumption data from each participated household. Figure 4.2 shows the results of power saving rate, in terms of *Power saving rate*, in the participated households[12]. The average power saving rate of the households in this experiment is –4.8%, which indicates an increase in energy consumption compared to the previous year. The range of power saving rate is from 27% and – 63%. However, about one-third households could save their electricity consumption. One quarter participated households successfully reduce, and the power saving rate of them is from 0% to 10%. While, there are few households

---

[12] One participant dropped out of the field experiment, reducing the total number of households to 52.

that save electricity by 10% or more (i.e., 5.8% of households save from 10% to 20%, and only 3.8% of households save more than 20%).

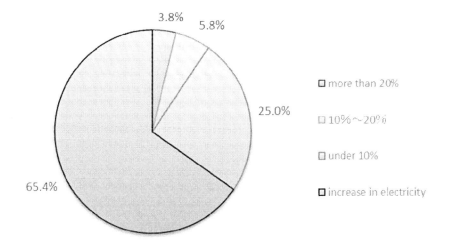

Figure 4.2. Results of power saving rate.

From the results, it can be said that the sampled Japanese households of our field experiment showed poor response to economic incentives, or the economic incentives offered in this experiment were possibly lower than the marginal cost of the electricity-saving behavior.

4.3. EMPIRICAL ANALYSIS

4.3.1. Statistical Test

If people respond to the economic incentives, their energy-saving behavior is expected to change. Here, we verify from the pre- and post-questionnaire data whether households' energy-saving behavior at home actually changed. We asked participant households about their energy-saving behavior prior to and after the questionnaire survey, such as the electricity-saving behavior of the representative, the encouragement by the representative to his/her family or by the family members to the rest of the family. We examine whether statistically significant differences in these behaviors were seen before and after.

We employed the following statistical test to examine the difference in each behavior before the experiment and during the experiment:

$$H_0: B_{before} - B_{under} = 0$$

B_{before} and B_{under} represent the power saving behavior before and during the experiment, respectively. If the above null hypothesis is rejected, we can assume that the economic incentive had a positive influence on the electricity-conservation behavior.

Table 4.2. Results of the statistical test

	before	after	difference	
Electricity-saving behavior of the representative	3.12	3.44	-0.33	**
Encouragement to electricity-saving behavior by the representative	2.75	2.88	-0.13	
Encouragement to electricity-saving behavior by the family	2.21	2.56	-0.35	

1. responses to Likert scale (always:5, usually:4, sometimes:3, rarely:2, never:1).
2. ***, **, * indicate the statistical significance at 1%, 5%, and 10% level.

Table 4.2 shows the results of the statistical test. We can confirm that only "electricity-saving behavior of the representative" is statistically significant difference, and the encouragement behaviors were not changed even with economic incentives. However, this simple statistical test did not take consider the socio-economic characteristics of each participated household. There may be household who are less likely to respond to, and household who are easily responsive to economic incentives. Then, using regression analysis, we will try to verify in the situation where socio-economic characteristics are controlled.

4.3.2. Econometric Analysis

This subsection examines whether an economic incentive encourages electricity-saving behavior among households. We used the following regression model to analyze the variables with *Power-saving rate* as the dependent variable:

$$Y_i = \alpha + \beta X_i + u_i \tag{4.2}$$

The dependent variable Yi indicates the *Power-saving rate* of the ith household. If the dependent variable takes a negative value, it implies that "households did not save electricity during our experiment." Xi indicates a vector of the independent variables, which includes age, gender of the household representative, his or her family size, number of children, whether living with parents, the representative's environmental awareness, change in the electricity-saving behavior of the representative, change in the encouragement by the representative to his/her family or by the family members to the rest of the family, size of home (area), average outdoor temperature, the number of air conditioners, and the number of televisions. We conducted pre-and post-questionnaire surveys to obtain these data of the independent variables. Table 4.3 presents the descriptive statistics of the above variables[13].

---

[13] The values of before and during for "Electricity saving behavior of representative," "Encouragement to do electricity-saving by representative" and "Encouragement to do electricity-saving by the family" are shown separately.

Kenichi Mizobuchi and Hisashi Tanizaki

Table 4.3. Descriptive statistics

Variable	Mean	Std. Dev.	Minimum	Maximum
Reduction (%)	-4.792	15.188	-63.170	26.910
Sex (men = 1, women = 0)	0.577	0.499	0.000	1.000
Age (respondent)	41.230	11.639	22.000	62.000
Gender (respondent) (male = 1, female = 0)	0.577	0.499	0.000	1.000
Family size	3.000	1.188	1.000	6.000
Number of children (under 12)	0.481	0.874	0.000	3.000
Living with parents (yes = 1, no = 0)	0.308	0.466	0.000	1.000
Environmental awareness *	4.077	0.518	2.000	5.000
Electricity saving behavior of respondent (before) *	3.115	0.704	2.000	5.000
Electricity saving behavior of respondent (during) *	3.442	0.777	2.000	5.000
Encouragement to do electricity-saving by respondent (before) *	2.750	1.064	1.000	5.000
Encouragement to do electricity-saving by respondent (during) *	2.885	1.041	1.000	5.000
Encouragement to do electricity-saving by the family (before) *	2.015	1.471	1.000	5.000
Encouragement to do electricity-saving by the family (during) *	2.558	1.092	1.000	5.000
Size of home (seven-grade by square meter)	4.154	1.764	1.000	7.000
Temperature (previous year - target year)	0.925	0.131	0.500	1.170
The number of air-conditioner	3.000	1.495	1.000	7.000
The number of TV	2.289	1.319	0.000	6.000

*Responses to Likert Scale (always = 5, usually = 4, sometimes = 3, rarely = 2, never = 1).

Eviews (version 7.0), which is an econometric software package, was used to estimate Equation (4.2). We use Whites' covariance matrix to adjust heteroscedasticity. Table 4.4 presents our estimation results. In order to examine the influence of change in behavior by incentive on households' power consumption, we focused on the following estimation coefficients: i) "electricity-saving behavior of the representative," ii) "the encouragement to engage in electricity saving offered by the representative," and iii) "the encouragement to engage in electricity saving offered by the family." We employed the following statistical test to examine the change in each behavior before the experiment and during the experiment:

$$H_0: \quad \hat{\beta} = 0$$

β represents the estimation coefficients for each variable, i.e., "electricity saving behavior of the representative (during – before)," "encouragement offered by the representative to engage in electricity saving (during – before)" and "encouragement offered by the rest of the family to engage in electricity saving (during – before)." If the above null hypothesis is rejected, it can be argued that changes in behavior by incentives have influenced electricity consumption.

According to the statistical test, the t-values (absolute value) of each variable were 0.312, 1.791, and 1.957, respectively. The 10 percent critical value of t-distribution with a degree of 38 yields a value of 1.686, leading to the following relationships: $0.312 < 1.686$, $1.791 > 1.686$, and $1.957 > 1.686$, respectively. As the null hypothesis is not rejected in the case of the first value, it is clear that the economic incentive did not introduce changes in the electricity-saving behavior of the representative. However, in the case of the second and third values - encouragement offered by the representative to engage in electricity saving – and - encouragement to engage in electricity saving offered by the family - the null hypothesis was rejected, confirming the positive effect of the economic incentive on electricity saving. This result is largely different from the result of Table 4.3. Since the power saving behavior is influenced by various socio-economic characteristics, it can be said that regression analysis that controls other factors is preferable to a simple difference test.

Kenichi Mizobuchi and Hisashi Tanizaki

Table 4.4. Estimation results

	Estimated coefficient	
Age	0.086	
	[0.270]	
Sex	-8.245	*
	[4.445]	
Family size	2.898	
	[2.527]	
Number of children	-6.894	*
	[3.582]	
Living with parents	-7.848	
	[6.986]	
Environmental awareness	5.263	
	[4.445]	
Electricity saving behavior of respondent	-0.842	
(during - before)	[2.695]	
Encouragement to do electricity-saving	4.629	*
by respondent (during - before)	[2.584]	
Encouragement to do electricity-saving	2.622	*
by the family (during - before)	[1.340]	
Size of home	-1.262	
	[2.026]	
Number of air-conditioner	2.335	
	[2.107]	
Number of TV	-1.328	
	[2.531]	
Temperature	-32.532	
	[19.051]	
Constant	-57.601	**
	[27.147]	
adjusted R square	0.315	
number of observation	52	

1. ***, **, * indicate the statistical significance at 1%, 5%, and 10% level.
2. The values in the parentheses indicate p-value of the estimated parameter.

As a reason why the change of the representative himself/herself did not lead to the energy saving behavior, the representative is a working individual who does not spend majority of his or her time at home on weekdays, unlike the other family members. Given that the unit of study is the household, behavioral changes in family members, who spend more time at home, are more prominent than those of the representative. This is a key finding of our study. None of the earlier studies on energy conservation have grouped members of a household according to their frequency of staying at home on a daily basis or identified a difference in behavior between members who stay at home and those who do not.

Our results show that the coefficient of "number of children under 12 years old" is negative and statistically significant, indicating a negative influence on electricity saving. The coefficient of "temperature" was also negative, but statistically insignificant. This can be attributed to the small difference in winter temperature between the study period and the previous year (only 0.6 Centigrade lower in the study period).

4.4. DISCUSSION

In this section, we investigate the main factor responsible for the negative average energy-saving rate: prior underestimation of the difficulty of energy-saving behavior (i.e., underestimation of the marginal cost of electricity saving). Figure 4.3 shows the participants' evaluation of the difficulties associated with electricity conservation before and after our field experiment. Before the experiment, majority of the respondents felt that the level of electricity conservation was "Neither easy nor difficult," and there were few households who responded that it was "Very difficult." However, after the experiment, 77% participants answered that it was "Difficult" or "Very difficult." Furthermore, the number of participants who responded "Neither easy nor difficult" was drastically reduced. The possible reason for such a change could be that the participants underestimated the difficulty of the power-saving behavior. Based on the results in Figure 4.3, most participants do not consider power-saving behavior to be difficult in the pre-questionnaire survey, and it turns out that they have realized the difficulty only after attempting to save energy. This implies that if researchers' enquiry about power-saving behavior through questionnaire alone without actually obtaining electricity usage data would provide biased results/would indicate a high possibility of obtaining different results than from the actual situation.

Moreover, the underestimation of such energy-saving behavior may indicate the high marginal cost of energy-saving behavior. Marginal cost is a unit of money (1% in this study) that represents the effort associated with saving energy. If the marginal cost of energy saving is less than the incentive, people should conserve electricity. However, as shown in Figure 4.2, most of the participants (65.4%) did not save electricity. This indicates that the marginal cost of power saving is substantial. Furthermore, the results in the figure indicate that people tend to underestimate the marginal cost of energy conservation. In such circumstances, even if the government asks the household to conserve electricity, not many households save electricity than predicted in advance. When asked the participating households in the post-questionnaire survey "Would you conserve electricity if the incentive amounts were doubled?," nearly 33% of the participants replied in the negative. This result implies that the marginal cost of saving electricity is very high. The underestimation of the marginal cost of energy-saving behavior of households has been verified in the research of Chapter 5.

Such underestimation of the marginal cost of energy-saving behavior provides an important implication for policymakers. When the government undertakes energy conservation measures with incentives for households, even if the government sets a target energy-saving rate and investigates the feasibility of such measures in advance, households' prior overestimation of the power-saving rate. Therefore, even if the government actually undertakes measures to save electricity, it may not be able to achieve the set target.

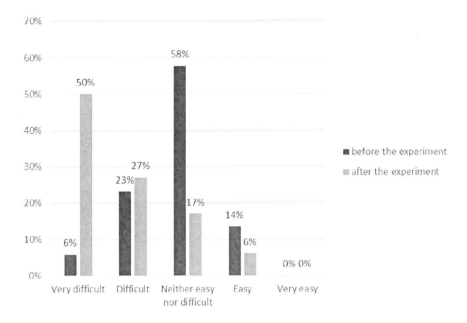

Figure 4.3. Changes in the difficulty of power saving.

4.5. LIMITATIONS

In this experiment, although crucial results were obtained, there were some limitations. The most important limitation is that a comparative analysis of the results was not possible among the groups, including the control group. In general, in field experiments, the treatment group (the group to which incentive, feedback, etc., were applicable) and control group (the group to which none of them were applicable) are randomly set. By comparing the results in each group, we can verify the influences of the treatment effect. Originally, this should be verified using nonparametric methods comparing the average values of each group. However, in this study, we abandoned this method owing to the small sample size (i.e., 52). Rather, a parametric analysis was performed with other factors controlled. However, for the comparison of our results with those of previous studies on economic incentives and other interventions such as information and feedback, setting of a control group may be necessary (Abrahamse et al. 2005). In the field experiments of Chapter 5, we compare the results by including the control group.

Moreover, in this study, the participants were not randomly extracted. In field experiments such as those conducted for energy saving, households with high consciousness of energy conservation and motivation may tend to take part in advance. To avoid the inclusion of participants who are highly motivated to conserve energy, we did not approach individual households to participate. Rather, for this study, participants were recruited at each workplace, and all employees at participating business sites participated. However, this method does not randomly extract participants. Most previous studies examining the effectiveness of incentives using data from master meters have similarly selected apartments and areas, instead of opting for random selection (Midden et al. 1983, McClelland and Cook, 1980, Petersen et al. 2007). However, in recent field experiments, it is necessary to randomly extract participants. Thus, this can be said to be one of the limitations of this study. In Chapter 6, by using the Propensity Score matching (PSM) method that creates pseudo randomization, the power-saving effect of households is verified.

CONCLUSION

In this chapter, we examined how economic incentives affect the energy-saving behavior of households, using field experiments. We conducted 3-week experiments for 53 households in Matsuyama city in Japan. Regarding the economic incentive, we designed it stepwise according to the power-saving rate. Experimental results revealed that economic incentives influence people's power-saving behavior to a certain extent (approximately 43% of the participating households succeeded in saving electricity). This finding is in agreement with that of previous studies (Winett et al. 1978, Midden et al. 1983, Petersen et al. 2007), which support the efficacy of the incentive. In particular, the results in Table 4.4 demonstrate that incentives also affected the behavior of the families who lived together. As Ek and Soderholm (2010) state, the traditional sampling method tends to attract participants who are highly motivated to conserve energy, leading to an upward bias in the results. In order to avoid such problems, in this study, we recruited participants from their workplace, so that all employees of the target workshop participate.

On the other hand, from the results of our experiment, the difficulty of saving electricity was clarified. Figure 4.2 shows that about two-thirds households did not conserve electricity. This was because the questionnaire results before and after the experiment showed that the participating households underestimated the marginal cost of energy-saving behavior (see Figure 4.3). Furthermore, based on the results of the questionnaire after the experiment, even if the incentive was doubled, 33% of the participating households said that they will not conserve electricity. This presents a valuable insight for policymakers targeting the household sector for reducing energy consumption and carbon dioxide emissions. On the premise that each house underestimates the marginal cost of energy saving, measures combined with methods other than incentives may be necessary. For example, target setting, feedback to inform of own power usage, and comparative feedback to compare usage with others may be effective. Chapter 5 examines the power-saving effect of countermeasures combining incentives and comparative feedback.

As also shown in the research in Chapter 3, in Japan, it is necessary to tackle the reduction of electricity consumption even for purposes other than carbon dioxide emissions reduction. Public power-saving pressure and the economic incentives have been found to have certain effects on energy conservation in the household sector; however, both have challenges on the length and size of the effect. To examine the policy mix that is a combination with some countermeasures, we will examine comparative feedback and the effect of replacing household appliances etc. in Chapters 5 and 6.

Chapter 5

FIELD EXPERIMENT 2: THE EFFECT OF CONTINUOUS REWARD: THE EFFECT OF ECONOMIC AND PSYCHOLOGICAL FACTORS ON ENERGY-SAVING BEHAVIOR

ABSTRACT

This chapter examines the impact of economic factors (incentive) and psychological factors (comparative feedback) on households' energy-saving behavior. Field experiments were conducted for 236 households in Matsuyama city, Ehime Prefecture, during the two months of October and November 2011. In this experiment, participating households were divided into three groups: i) Reward; ii) Reward with Comparative Feedback; and iii) Control. Thereafter, the average power-saving rate of each group was compared. The results suggest that the first two groups had a significantly higher power-saving rate than the third group. In addition, from the result of the quantitative analysis, it was found that social norm significantly promotes power-saving behavior. In other words, people tend to demonstrate energy-saving behavior when others around them do so. Furthermore, as in Chapter 4, from the results of the preliminary questionnaire and post-questionnaire results, it was found that households underestimated the marginal cost of energy-saving behavior before they began to save electricity (this chapter is a revision of Mizobuchi and Takeushi, 2013).

Keywords: electricity saving, reward, comparative feedback, field experiment

5.1. INTRODUCTION

In this chapter, we evaluate the influences of both economic (incentive) and psychological (comparative feedback) factors on power conservation behavior among households. We conducted a field experiment based on an intervention study from October to November of 2011 in randomly selected Japanese households. To the best of our knowledge, this is the first study that examines the influence of both factors on power-saving behavior based on behavioral data (Jeroen 2008). Moreover, we divide the decision of power-saving behavior into *whether to save* and *how much to save*, and reveal the main contributing factors to each type of decision-making.

This paper is organized as follows. Section 5.2 describes our field experiment, in which we used subsidies as an economic incentive to promote electricity conservation behavior within households. Section 5.3 presents our empirical analysis. Section 5.4 discusses the marginal costs of electricity conservation behavior. Section 5.5 presents the conclusion.

5.2. EXPERIMENT

5.2.1. Set Up

Our field experiment lasted for 8 weeks, from October to November 2011. Each enrolled household was entitled to receive a reward for reducing its electricity consumption to a level below their previous year's consumption for the same period. Electricity consumption was measured in kWh to ensure comparability across households, regardless of the energy fuels used. The reduction of electricity consumption (*Reduction*) was calculated as follows:

$$Reduction(\%) = \frac{X_p - X_c}{X_c} \times 100, \tag{5.1}$$

where X_c and X_p represent the electricity consumption during the experiment and during the same period of the previous year, respectively. If *Reduction* was larger than 1% (*Reduction* $\geq$ 1), households could receive a reward (economic incentive). The amount of the reward was 200 yen (about $2) per 1% of electricity consumption reduction (We have assumed the following exchange rate: $1 = 100 yen). For example, if actual *Reduction* was 10%, households were paid 2,000 yen.

On the other hand, we also examined the effectiveness of comparative feedback, in which each household received information about the average amount of electricity saving by other households (e.g., the value of *Reduction*). We conducted the comparative feedback twice: the first one informed households of the target levels of electricity saving by means of pre-experiment questionnaires in late September, and the second one provided information on the values of *Reduction* from the first month (i.e., October) in early November.[14]

To examine the effectiveness of economic incentives and of comparative feedback, households were assigned randomly to three groups: (1) Reward; (2) Reward with Comparative Feedback; and (3) Control. We use a non-parametric approach to compare the three groups

---

[14] The first comparative feedback is not a real value of *Reduction*, and there may be a difference between the target value and real values. However, participants can compare others' target reduction levels with their own. Therefore, in terms of comparison with other participants, we deal with it as the first comparative feedback.

in Section 5.2.3. To avoid the problems of sampling faced by previous studies, we set the sample size of every group as larger than 50. Moreover, we also use a parametric approach—that is, a regression analysis—taking *Reduction* as the dependent variable to examine the influences of economic and psychological factors on electricity conservation behavior (see Section 5.3 for details).

We conducted two questionnaire surveys, one before and one after the field experiment. In the pre-experiment questionnaire survey, we asked participants about the following aspects: attitude toward environmental issues, level of environmental concern, attitude toward electricity conservation, intended target level of electricity saving as well as level of confidence in meeting this target, and the types of home electrical appliances in the household. Finally, we elicited social norm, socio-economic, and demographic details. In our study, the degree of social norm is measured by the question "Do you consider how often people close to you are attempting to save electricity/demonstrate electricity-saving behavior?" Comparison with neighbors or other close persons will construct people's social norms and will induce them to conserve energy (Nolan et al. 2008 and Allcott 2011). Responses were measured on a five-point Likert scale ("Always" = 5; "Usually" = 4; "Sometimes = 3; "Rarely" = 2; "Never" = 1), and high-score responses indicate a high level of attention to social norms (and vice versa). In the post-experiment questionnaire survey, we asked participants, *inter alia*, about changes in their attitudes toward electricity conservation, the challenges in conserving electricity, and the appropriateness of economic incentives.

5.2.2. Respondents

Our field experiment was conducted in Matsuyama city in Ehime Prefecture that has approximately 480,000 inhabitants and is located in

Western Japan. In August 2011, 1,000 households in Matsuyama city were randomly selected and were sent letters informing them about the study and inviting them to participate in our field experiment. Aside from a broad description—that is, the study was about an experiment in energy saving—we did not divulge any details about the field experiment in the initial invitation letter. Of the households invited to participate in this study, 236 (13.1%) returned a completed pre-experiment questionnaire. Compared to a national sample (Population Statistics, 2009), male respondents were overrepresented, [15] and respondents in the age category of 55–69 years were slightly underrepresented.

As described earlier, participating households were entitled to receive subsidies based on their *Reduction*, which was calculated according to Equation (1). Participants could access their historical data on electricity consumption per month by logging onto the Shikoku Electric Power website, using their unique identification numbers. We collected these consumption data for the previous year from the households and used them as the measurement criteria. Additionally, for tracking ongoing consumption, every household was asked to send a copy of its monthly electricity consumption record.

5.2.3. Non-Parametric Analysis

Figure 5.1 shows the results of electricity conservation, in terms of *Reduction*, in the sampled households at the end of eight weeks. About 70% of the households successfully reduced their electricity consumption. There are some possibilities as causes for this result: the households showed a high response to economic incentives or comparative feedback; the economic incentives offered in this experiment were perhaps higher than the marginal cost of the

---

[15] Most respondents were householders. This might be the reason for male overrepresentation.

electricity-saving behavior; or the temperature of the target year may have been cooler than that of the previous year. As Figure 5.1 shows, in particular, the greatest percentage of households (18.8%) was in the 5%–10% *Reduction* bracket, followed by the 10%–15% bracket (17.3%), and the 1%–5% (15.4%). However, the percentage of households that saved more than 15% was small, and it decreased as the values of *Reduction* increased. Additionally, about 28.8% of the households did not conserve electricity; therefore, we can confirm that a certain percentage of households did not respond well to economic incentives and comparative feedback.

The average *Reduction* of the households in this experiment is 5.4%, indicating a decrease in energy consumption compared to the previous year. Here, we investigate the effects of interventions— economic incentive and comparative feedback—statistically.

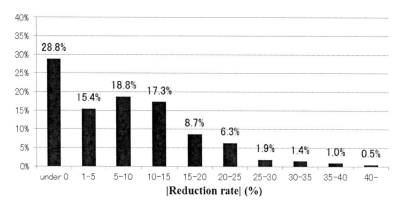

Figure 5.1. Power conservation of participants.

Table 5.1 shows the average electricity-saving rate (i.e., average *Reduction*) of the three groups: (i) Reward; (ii) Reward with Feedback; and (iii) Control. The average saving rate is highest in the Reward with Feedback group (8.2%), followed by the Reward group (5.9%), and the Control group (1.6%). Because our study aims to investigate the effects of intervention, we test the following null hypothesis.

$$H_0 : \gamma_i - \gamma_j = 0 \qquad i \neq j, \quad i, j =$$ (i), (ii), (iii),

where γ_i indicates the average *Reduction* of *i*th groups. Table 5.2 shows the t-statistics of three hypotheses—that is, Reward vs. Control, Reward with Feedback vs. Control, and Reward vs. Reward with Feedback. From the t-statistics of the first hypothesis, Reward vs. Control, the null hypothesis was rejected at the 10% significance level (1.77 > 1.65, p = 0.1); thus, the effect of the economic incentive could be found statistically. Moreover, from the t-statistics of the second hypothesis, Reward with Feedback vs. Control, the null hypothesis was also rejected at the 5% significance level (2.65 > 1.98, p = 0.05); this result enhances the effect of the economic incentive. However, the intrinsic effect of comparative feedback was obscure, because the null hypothesis was not rejected (1.10 < 1.65, p = 0.1) for the third hypothesis, Reward vs. Reward with Feedback. This result is inconsistent with those of Midden et al. (1983) and Brandon and Lewis (1999), which supported the effectiveness of only comparative feedback, but is consistent with those of McClelland and Cook (1980) and Petersen et al. (2007), which supported the effectiveness of comparative feedback with reward.

Table 5.1. A comparative analysis of average electricity-saving rate of each group

	(i) Reward	(ii) Reward + Feedback	(iii) Control
Average	-5.9	-8.2	-1.6
Variance	207.23	129.58	207.00
Number of obs.	103	53	52
	(i) vs (iii)	(ii) vs (iii)	(i) vs (ii)
t-value	-1.77*	-2.65**	-1.10

* and ** indicate 10% and 5% level of statistical significance, respectively.

5.3. PARAMETRIC ANALYSIS

This section presents the results of a parametric analysis (i.e., regression analysis) undertaken to examine whether economic and psychological factors affect electricity-saving behavior within households.

5.3.1. Model and Data

We begin with a Tobit model for analyzing the variables with *Reduction* as the dependent variable because there are a large number of households that do not save electricity:

$$Y_i^* = \alpha + \beta X_i + \varepsilon_i$$
$$Y_i = \max(0, Y_i^*) \qquad i = 1, 2, ..., N \tag{5.2}$$

The dependent variable Y_i indicates the *Reduction* of the ith household and is censored at zero. Households with a negative value of *Reduction* in our experiment are treated as households "not saving electricity right from the beginning" or "having abandoned electricity conservation sometime during the experiment." X is a vector of the independent variables: a comparative feedback dummy, a reward (i.e., economic incentive) dummy, an anticipation of the electricity-saving behavior of individuals residing nearby the household (i.e., social norm), the New Ecological paradigm (NEP) score, target level of *Reduction* shown in the pre-questionnaire, gender and age of the household representative, income level, family size, type of homeownership, size of home (area), and cooling degree days [16]

---

[16] The difference between an average temperature higher than 24 °C and 22 °C. The day of meter reading by the power company is different for each household. Therefore, we summed up the differences for the experimental term and set an independent variable.

(previous year – target year). Data on the independent variables were obtained from the pre- and post-study questionnaires. Table 3 presents a statistical summary of the above and dependent variables.

Table 5.2. Descriptive statistics

Variable	Mean	S.D.	Min	Max
Reduction (%)	5.413	13.83	-69.000	42.800
Psychological factors				
Comparative feedback with Reward (yes = 1)	0.255	0.437	0.000	1.000
Social norm	3.197	0.663	0.000	5.000
(Individuals residing nearby attempt electricity-saving behavior)*1				
NEP-score	57.091	6.896	32.000	71.000
Target *2	3.413	1.338	1.000	8.000
Economic incentive				
Reward without feedback (yes = 1)	0.495	0.501	0.000	1.000
Socio-economic variables				
Gender (household representative) (male = 1)	0.654	0.477	0.000	1.000
Age (household representative)	42.6.9	10.802	21.000	80.000
Income *3	1.894	1.098	1.000	6.000
Family size	3.135	1.228	1.000	6.000
Homeownership (yes = 1)	0.668	0.472	0.000	1.000
Size of home *4	3.74	1.448	1.000	7.000
Weather variable				
Cooling degree days (previous year - target year)	8.761	7.228	-0.700	27.800

*1: Responses on a five-point Likert scale (Always = 5; Usually = 4; Sometimes = 3; Rarely = 2; Never = 1).

*2: Over 30% = 8; 25–30% = 7; 20–25% = 6; 15–20% = 5; 10–15% = 4; 5–10% = 3; 1–5% = 2; under 0% = 1.

*3: Over 15million yen = 6; 10–15million yen = 5; 8–10million yen = 4; 6–8million yen = 3; 4–6million yen = 2; under 4 million yen = 1.

*4: The number of rooms (over seven rooms = 7; six rooms = 6; five rooms = 5; four rooms = 4; three rooms = 3; two rooms = 2; one room = 1).

5.3.2. Estimation Results

The estimation results from the Tobit model are reported in the first column of Table 5.3. The *comparative feedback* and *reward* have a positive effect on electricity-saving behavior; both variables have coefficients that are positive and statistically significant. In the previous section, we showed the efficacy of economic incentives and comparative feedback against the control group by a non-parametric approach. Here, both the values of the estimation coefficient were statistically significant; therefore, we confirmed by the parametric approach that reward and comparative feedback (with reward) have an effect of encouraging electricity-saving behavior. However, the comparison of estimated coefficients between the *comparative feedback* and *reward* is not statistically significant. Thus, similar to the result in Section 5.2.3, the efficacy of comparative feedback (without reward) is obscure.

The coefficient of the social norm is positive and statistically significant, showing that others' behavior does have an impact on one's own electricity-saving behavior. This impact could stem from two sources. First, the overall household behavior in electricity saving may influence people's views concerning personal responsibility. Specifically, if others participate in electricity saving, the individual may experience loss in his or her self-image as a morally responsible person if he or she does not do so. Second, the presence of explicit prescriptive social norms stemming from, for instance, family members and close friends also has an effect. This influence may well constitute a complement to the above perceptions of others' contributions. If close friends and family stress the importance of saving electricity, an individual may deduce that others also believe it is important. Nyborg et al. (2006) presented a theoretical framework in which individual responsibility for pro-environmental behavior depends on beliefs about others' behavior. Using household survey data, Ek and Soderholm

(2008) showed the positive influence of social norms on participation in the Swedish green-electricity market. Furthermore, Nolan et al. (2008) and Allcott (2011) showed the positive influence of social norms on U.S. household energy-saving behavior.

Two other psychological factors, such as the NEP score and target, have no influence on electricity conservation behavior; both coefficients are statistically insignificant. The positive effect of the NEP score has manifested in other pro-environmental behaviors, such as participation in the green-electricity market, in some previous studies (Clark et al. 2003; Kotchen and Moore 2007; Ek and Soderholm 2008). However, electricity-saving behavior not only has a public good aspect—that is, being pro-environmental—but also a private good aspect, such as saving one's own money. This consideration may be the cause of the insignificant effect of the NEP score. Another possible reason is the *Motivation Crowding Theory*, or the *crowding-out effect*, which suggests that external intervention via monetary incentives may undermine intrinsic motivation (Deci 1971). This effect is present widely in the economy and society (Deci et al. 1999; Frey and Jegen 2001). If the individuals affected perceive external interventions to be *controlling*, both self-determination and self-esteem suffer, and such individuals react by reducing their intrinsic motivation in the activity controlled. From this effect, participants who save electricity because of motivation from intrinsically high environmental concern might experience reduced motivation to save electricity if monetary rewards are introduced, and this influence might be statistically insignificant regarding the NEP parameter.

The coefficient on income is statistically insignificant, and this result is a little different than what was expected. In the theoretical model of public goods, the coefficient of household income will become positive, and this outcome has been supported empirically (Kotchen and Moore 2007), as pro-environmental behaviors do, in fact, increase with income. However, similar to the result of the NEP score above, electricity-saving behavior does not have the nature of a pure public good, thus perhaps causing this result. Family size and homeownership have a negative and statistically significant effect on electricity saving, suggesting the importance of considering disposable income. The size of home is positive and statistically significant. This result may indicate the possibility that there is large electricity-saving potential in a house that has many rooms. Finally, the coefficient of the cooling degree day (previous year–target year) is statistically significant, and the positive sign indicates that if the target year becomes cooler than the previous year, the amount of time the air conditioner is used decreases, a situation conducive to electricity saving. In reality, the average temperature within the experiment period was lower by 0.9°C than it was the previous year.[17]

A feature of the Tobit model is the restriction that explanatory variables are assumed to influence the extensive and intensive margins of electricity-saving behavior in the same way. That is, an implicit assumption is made that the decision of *whether to save electricity* is the same as the decision of *how much to save*. However, it is possible that the explanatory variables influence electricity saving on the extensive and intensive margins in different ways. Smith et al. (1995) and Kotchen and Moore (2007) make this observation and find empirical support for it in a study of charitable contributions to a rural healthcare facility and a green-electricity program, respectively. We explore the same possibility here by decomposing the Tobit model into a probit model for the decision of *whether to save electricity*, and a truncated regression model for the decision of *how much to conserve*.[18]

---

[17] In the post-study questionnaire, we asked participants to indicate a factor that was important for electricity saving. Thirty-five percent households felt that *cooperation from family members* was the most important factor. *Willingness to act* was identified as an important factor by 31%, while 15% households believed that it was *temperature*. Only a small percentage of respondents identified *information on energy saving* (4%) as an important factor.

[18] Similar to Kotchen and Moore (2007), our sample includes all households that participated in our field experiment; therefore, the analysis of the intensive margin needs no correction for sample-selection bias. For such cases, Greene (2008) notes that the appropriate decomposition of a Tobit model is into a probit model and a truncated regression model.

Table 5.3. Estimation results

Variable	Model			
	(1)	(2)	(3)	
	Tobit	Probit	Truncated regression	
Psychological factors				
Comparative feedback with Reward	6.849 ***	0.867 ***	2.351	
	[2.353]	[0.288]	[2.756]	
Social norm	2.934 **	0.310 *	2.356	*
	[1.178]	[0.163]	[1.415]	
NEP-score	-0.015	0.017	-0.175	
	[0.112]	[0.015]	[0.130]	
Target (eight-grade by Reduction (%))	0.446	-0.042	1.471	**
	[0.706]	[0.074]	[0.728]	
Economic incentive				
Reward (yes = 1, no = 0)	3.868 *	0.342	2.227	
	[2.098]	[0.240]	[2.460]	
Socio-economic variables				
Gender	3.009	0.230	3.069	
	[1.865]	[0.230]	[2.197]	
Age (household representative)	-0.062	-0.004	-0.082	
	[0.087]	[0.010]	[0.097]	
Income (six-grade by yen)	-0.001	-0.045	0.440	
	[0.756]	[0.097]	[0.911]	
Family size	-2.440 ***	-0.090	-3.487	***
	[0.832]	[0.096]	[1.037]	
Homeownership	-5.135 **	-0.554 *	-2.656	
	[2.259]	[0.292]	[2.627]	
Size of home	2.003 ***	0.227 **	1.288	
	[0.731]	[0.099]	[0.901]	
Weather variable				
Cooling degree days	0.234 **	0.009	0.288	**
	[0.113]	[0.014]	[0.116]	
Constant	-5.142	-1.803	10.580	
	[8.821]	[1.148]	[10.695]	
Number of observations	208	208	146	
S. E. regression	8.388	0.450	7.630	
Log likelihood	-606.983	-116.937	-476.116	

1. ***, **, and *, represent statistical significance at 1%, 5%, and 10% level, respectively.
2. Standard deviation in parentheses.

We report the results of these two models in the second and third columns of Table 5.3. The qualitative results of the probit model are similar to those of the Tobit model; most of the coefficients have the same sign except for the NEP score and target. However, the levels of statistical significance are substantially different. Thus, the variables that influence only the extensive margin of electricity saving (*whether to save electricity*) are different from those that jointly influence the extensive and intensive margins (*how much to save*). Comparative feedback, social norm, and size of home have a positive and statistically significant effect on the decision of *whether to save electricity*, whereas homeownership has a negative and statistically significant effect. Here, the marginal effect of comparative feedback is statistically larger than that of reward. Thus, under the existing economic incentives, if households know the amount of other households' electricity saving and can compare energy-saving levels under the economic incentives, they may be motivated to compete and may conserve electricity consumption more than without the comparative feedback (McClelland and Cook 1980; Petersen et al. 2007).

The results also differ substantially when we focus on the intensive margin (*how much to save*). In the truncated regression model, two psychological factors, one socio-economic factor, and the weather variable are statistically significant explanatory variables in the decision of *how much to save*, and economic incentive factors have no effect. In particular, the coefficient of *target*, which was not statistically significant in both the Tobit and probit models, is statistically significant. Becker (1978) showed that setting a relatively difficult goal appeared to be more effective in reducing energy use than setting a relatively easy goal. Our result of the truncated regression model is consistent with his result.

These results provide evidence that the decision of *whether to save electricity* is not determined in the same way as the decision of *how much to save*. Specifically, we found that comparative feedback, social norm, homeownership, and size of home influence the decision about *whether to save electricity*, and social norm, target setting, family size, and cooling degree days influence the decision of *how much to save*.

5.3.3. The Effects of Economic and Psychological Factors

Which is more important in terms of electricity conservation behavior: economic factors or psychological factors? Our results show that both economic and psychological factors are important. As we showed in subsection 5.3.2, however, the effectiveness of these two factors differs based on the timing of the decisions, in terms of the effect that each factor has on electricity-saving behavior—that is, *whether to save electricity* and *how much to save*. From the estimation result of the probit model in the second column of Table 5.3, the coefficients of both comparative feedback (with reward) and social norm are positive with a statistically significant effect. Therefore, in the decision of *whether to save electricity*, both economic and psychological factors would affect electricity-saving behavior. Here, the marginal effect of comparative feedback (with a reward) is larger than that of the social norm. Therefore, in the decision of *whether to save electricity*, we might conclude that comparative feedback with a reward will become the most positive decisive factor. From the estimation result of the truncated regression model in the third column of Table 5.3, on the other hand, only psychological factors—that is, *social norm* and *target*—are statistically significant. Thus, in decisions concerning *how much to save*, psychological factors would affect electricity-saving behavior the most. Here, only social norm has a positive and statistically significant effect on both decisions of *whether* and *how much*, which imply that social norms do have a consistent positive impact on one's own electricity-saving behavior.

5.4. DISCUSSION AND LIMITATIONS

This section discusses the two major reasons for the 28.8% of participants who demonstrated negative *Reduction* in this study: (1) lack of understanding of the difficulties in electricity conservation; and (2) the marginal cost of electricity saving. Figure 5.2 shows the participants' evaluation of the difficulties associated with electricity conservation before and after the experiment. Before the experiment, the majority of the respondents felt that the level of electricity conservation was "Neither easy nor difficult." However, at the end of the experiment, 54% of the participants felt that saving electricity was "Difficult" or "Very difficult." This change in response indicates a limited initial understanding of the challenges associated with the conservation behavior. This result may also suggest that the pre-experiment responses concerning electricity-saving behavior have an upward bias, as participants expect to save more electricity before actually acting on it. In other words, using only a stated intention to evaluate household energy saving may not be appropriate, given the likelihood for overestimation.

The second reason is that the marginal cost of energy saving is greater than the incentive amount. As described earlier, most of the households underestimated the difficulties in conservation before the experiment; that is, they underestimated the marginal cost of energy saving. However, actually making efforts to reduce consumption may have helped them realize the greater marginal cost of conservation, thus resulting in cessation of conservation efforts. To the question "Would you conserve electricity if the incentive amounts were doubled?" in the post-experiment survey, nearly 71% households that failed in reducing their consumption returned negative responses. This result also supports the notion that the marginal cost of saving electricity is considerable.

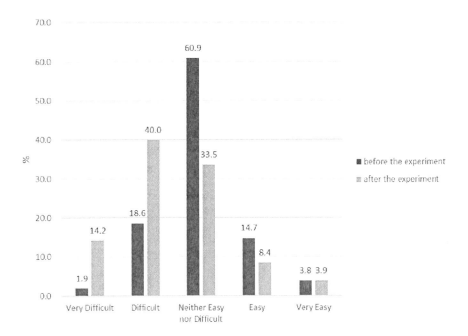

Figure 5.2. The evaluation of difficulty of electricity conservation behavior.

These results carry important implications for policymakers targeting energy conservation in the household sector: before introducing incentives to promote conservation, it is necessary to analyze the marginal cost of electricity saving and to ensure that the incentives exceed this cost. Moreover, it would be worthwhile to combine economic incentives with other interventions aimed at lowering the marginal cost. For example, as we showed in Section 5.3, the comparative feedback under economic incentives is very effective in the decision of whether to save electricity.

CONCLUSION

Through an eight-week field experiment, this study investigated the influences of economic and psychological factors on electricity conservation behavior among Japanese households. A few studies have examined the influence of the two factors on pro-environmental behavior (Clark et al. 2003; Jeroen 2008; Kotchen and Moore 2007); however, no studies solely focus on the electricity-saving behavior with real data. From the results detailed in Sections 5.2 and 5.3, we have shown the effectiveness of both economic and psychological factors in encouraging electricity conservation. However, the effectiveness of these two factors is divided, based on the timing of the decisions, in terms of the influence of each factor on electricity-saving behavior— that is, *whether to save electricity* and *how much to save*. Economic incentives are more effective than psychological factors such as social norms, environmental concerns, and target setting regarding decisions about *whether to save electricity*, especially under comparative feedback. On the other hand, in decisions about *how much to save electricity*, psychological factors such as social norms and target setting are effective. Only social norm has a consistent positive effect on both *whether to save* and *how much to save*. The effectiveness of economic incentives on electricity saving is in agreement with that of previous studies (Midden et al. 1983; Petersen et al. 2007; Winett et al. 1978). The effects of psychological factors—such as comparative feedback, social norm, and target setting—are also consistent with results from previous studies (Becker 1978; Midden et al. 1983; Brandon and Lewis 1999; Ek and Soderholm 2008; Nolan et al. 2008; Allcott 2011). However, this study does not suffer from the drawbacks identified in some earlier works, including sampling issues and small sample size. Our experimental households were randomly selected, and the sample size of 236 households is higher than that of previous studies, ensuring high statistical reliability.

Underestimation of the marginal cost of saving electricity is one of the reasons that led about 30% of the participants to cease their energy-saving efforts. This discovery presents a valuable insight for policymakers who target the household sector in reducing carbon dioxide emissions. Before offering economic incentives to promote energy conservation, the marginal costs of electricity conservation should be evaluated. If the costs are substantially high, economic incentives should be combined with other interventions, such as tailored information and goal setting, to lower the marginal costs. Comparative feedback, as examined in our study, might be a potentially useful intervention in this regard.

Chapter 6

FIELD EXPERIMENT 3: THE EFFECT OF ENERGY-SAVING INVESTMENT: THE ELECTRICITY-SAVING EFFECT FROM THE REPLACEMENT OF AIR CONDITIONER

ABSTRACT

In this chapter, we examine the magnitude of power-saving effect when households replace energy-consuming air conditioners with energy-saving ones. We obtained monthly electricity usage data for two years from 2013 and 2014 from 733 households living in the Kansai area in Japan. Furthermore, we conducted a questionnaire survey on household attributes and home appliances ownership status among the participating households. Based on the results of the questionnaire survey, we divided the households into household groups that had replaced energy-consuming air conditioners with energy-efficient ones during the survey period, and those that did not do so. Thereafter, we compared the average electricity usage of each group. We adopted a method combining the propensity score matching (PSM) and the difference-in-differences (DD) methods in order to control household attributes and others. We found no significant power-saving effect by such replacement.

Keywords: energy saving investment, air conditioner, household, propensity score matching, difference-in-differences

6.1. INTRODUCTION

While central cooling and heating systems are popular in European and American countries, unit cooling and heating, where an air conditioner fixed in a room provides heating or cooling only to the specific room, are popular in Japan[19]. Therefore, Japanese households make more frequent purchase and replacement decisions for air conditioners than European and American households.

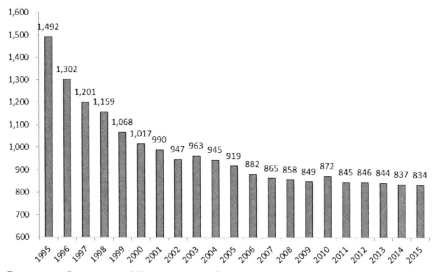

Data source: Resources and Energy Agency, Japan.

Figure 6.1. Term power consumption of air conditioner.

---

[19] According to EDMC (2015), room air conditioners have a diffusion rate of 90.6% in Japan and their average number per household is 2.76 based on 2013 data.

In Japan, air conditioners account for 60% of household electricity consumption during peak summer and winter hours (Ministry of Economy, Trade and Industry, 2013), and thus, we can anticipate large power saving by the replacement of non-energy-efficient air conditioners with energy-efficient ones. Figure 1 shows that there was substantial improvement in energy efficiency (44.1%) of air conditioners sold between 1995 and 2015 in Japan[20]. Moreover, as the rate of improvement between 2005 and 2015 has not been large (9.2%), we can say that it has reached an equilibrium.

In general, for evaluating the energy conservation effect of an energy-efficient investment, the randomized control trial (hereafter, RCT) is considered an ideal method. The validation the methodology of Dubin et al. (1986), Davis (2008), and Davis et al. (2014) is based on the RCT, and thus, their results are considered trustworthy. However, in the field of social psychology, the RCT method is not considered fully reliable, because it has the potential to include the Hawthorne effect. In concrete terms, under the RCT, a select group of people may be conscious of them being observed and this feeling could distort their usual behavior. For example, for evaluating a given educational program, a classroom is selected randomly, and the stakeholders, including the school management representatives, teaching staff, and parents or guardians of the children, visit the selected classroom. The presence of these visitors may make the students of the class conscious and thus motivate them to increase their efforts to study than is usual.

---

[20] The targets are some newly introduced representative energy-efficient air conditioners in the reference year that have both cooling and heating functions, with a cooling capacity of 2.8kW. They are assumed to be used in typical wooden houses in Tokyo area. The assumed usage time is 18 hours from 6 a.m. to 12 a.m. The assumed usage period of cooling is from June 2nd to September 21st, and the preset temperature is 27 °C. The assumed usage period of heating is from October 28th to April 14th, and the preset temperature is 20°C.

This phenomenon is called the Hawthorne effect. In this case, even if there is no direct effect of this educational program, a significant effect of the program may erroneously result due to the Hawthorne effect. Similarly, households who participated in the RCT (who were selected by this program, received energy-efficient appliances at no cost, and reported their electricity consumption data) may be conscious of being observed and hence may behave differently than usual (e.g., they attempt to save power).

This study examines the existence of the causality effect of power saving by replacement of energy-efficient air conditioners. However, we do not use the RCT method. We compared the actual electricity consumption data of two household groups, one being the treatment group that replaced their air conditioners with energy-efficient ones in the previous two years, and another being the control group that did not do so. In our study, the allocation between the treatment group and the control group was not random, and therefore, it is possible that several socio-economic characteristics affect the replacement behavior. Towards this, we employ the propensity score matching (PSM) method to adjust the covariates (i.e., socio-economic characteristics) of both treatment and control groups. Moreover, to exclude the unobservable factors, we combined the difference-in-differences (DD) method with the PSM method and estimated the power-saving causality effect by the replacement of air conditioners.

This section is organized as follows: Section 6.2 explains our analytical method combining both the PSM and DD methods. In Section 6.3, the data of empirical analysis are described. Our study targeted Japanese households who live in Kansai area, and we conducted a web-based questionnaire survey on them. Moreover, we asked them to exhibit their electricity consumption data for the previous two years. Section 6.4 presents the results of our empirical analysis and discusses them. Section 6.5 presents our conclusion.

6.2. ANALYTICAL METHOD

6.2.1. Potential Outcome Model

To investigate the power-saving effect by replacing an energy-consuming air conditioner with an energy-efficient one, our study uses the method developed by Roy (1951) and Rubin (1974), which has been used widely for estimating the causality effect. Here, a binary treatment indicator Di equals 1 if household i switched to the energy-efficient air conditioner and 0 otherwise. If we let Yi(Di) be the potential outcome of the replacement decision for household i, the treatment effect of household i may then be written as follows:

$$\delta_i = Y_i(1) - Y_i(0) \tag{6.1}$$

Here, for each i, Yi(1) and Yi(0) are counterfactual, and only either of them can be observed (fundamental problem of causal inference). If we assume the homogeneity of each household of a population, we have Yi(1) = Yj(1) and Yi(0) = Yj(0). Therefore, for each household i and j in a population, we have the following:

$$\delta_i = Y_i(1) - Y_i(0) = Y_j(1) - Y_j(0) = Y_i(1) - Y_j(0)$$

Further, we can estimate the treatment effect at the individual household level. That is, we consider only the difference between the potential outcome of households in the treatment group and that of those in the control group. However, as pointed out by Xie (2013), there is always an issue of heterogeneity with regard to individuals or households in the area of social science. Therefore, it is unrealistic for the analysis to assume homogeneity for each household, and it is difficult to adjust the estimation result to reality. For this problem, another widely used method that considers aggregate-level information, as opposed to the individual-level one, is the average treatment effect (hereafter, ATE):

$$\delta_{ATE} = E[Y(1) - Y(0)] \qquad (6.2)$$

Moreover, the average treatment effect on the treated (hereafter, ATT) method is widely used (i.e., the power-saving effect of the replacement on households that opted for it):

$$\delta_{ATT} = E[\delta / D = 1] = E[Y(1) / D = 1] - E[Y(0) / D = 1] \qquad (6.3)$$

In this study, we employ the ATT and estimate the power-saving effect of the replacement.

Here, if the assignment of the replacement was random (i.e., RCT), $E[Y(0) / D = 1] = E[Y(0) / D = 0]$ and δ_{ATT} would be identical. Further, the mean outcome (power consumption) of households that did not replace would serve as the counterfactual outcome of replacement households. However, if the assignment was not random (e.g., in the case of an observational study), the estimates of δ_{ATT} may suffer from a selection bias. That is, observable and unobservable household characteristics, which affect the decision to switch to energy-efficient air conditioners, also affect the electricity demand.

In the case of a non-random assignment, our identification of δ_{ATT} relies on two standard assumptions. First is the conditional independence assumption (hereafter, CIA), which means that conditional on the set of relevant covariates, the assignment of the treatment is independent of the potential outcome[21]. Second is the assumption of the common support (or overlap). This assumption

---

[21] That is, although the assignment of treatment can be dependent on the observable covariates, if we control for these covariates, we can rightly consider that the assignment of treatment was determined almost randomly without relating to the observable covariates.

implies that households with the same covariates have a positive probability of both replacement and non-replacement. In other words, each household has a positive probability of being in the treatment ("replacement") group and the control ("non-replacement") group.

This study uses the matching technique to estimate the ATT. For each household in the replacement group, this method employs a household from the control group, which has similar covariates with that of the replacement group, as the counterfactual. The difference between the power consumption of the household in the replacement group and that of the household in the control group may then be attributed to the replacement with energy-efficient air conditioner. That is, matching mimics "randomization" by balancing the distributions of the relevant characteristics (covariates) in the treatment group and the control group to attain independence between a household's decision to replace with an energy-efficient air conditioner and its decision to save power.

6.2.2. Propensity Score Matching

Rosenbaum and Rubin (1983) defined the probability of the treatment indicator variable that is conditional on the observable covariates $P(D_i|X_i)$ as the propensity score. To match each household between the treatment and control groups based on this propensity score, we can maintain independence between the decision of replacement (assignment) and the decision of power saving (potential outcome).

In the actual analysis, we cannot know the value of propensity score preliminarily, and thus, we need to estimate it. First, we estimate the probit (or logit) model that regresses the decision of replacement on a set of relevant observable covariates. Then, based on the estimation coefficients, the probability of the replacement (the propensity score) is predicted for each household i. These propensity scores are then used to identify households in the control group that best match those in the replacement group. In this sense, the propensity score aggregates the information in the relevant covariates into a single index. Therefore, under the PSM method, households in the replacement group may be paired with those in the control group exhibiting very different covariate values but close propensity scores. The ATT is estimated by calculating the difference between the power consumption in the replacement group households and their matches in the control group as follows:

$$\delta_{ATT}^{PSM} = E_{CP}\{E[Y_i(1)/D_i = 1, P(D_i/X_i)] - E[Y_i(0)/D_i = 0, P(D_i/X_i)]$$

(6.4)

Here, CP indicates the common support, which means an overlapping interval between the propensity scores of replacement households and those of control group households.

This study employs three types of common propensity score matching—nearest neighbor matching (NNM), radius matching, and kernel matching—where one household from the replacement group is paired with one household from the control group. Usually, there is a trade-off between bias and variance in each matching method. That is, if we reduce the variance, the bias of estimate would increase (vice versa). In the NNM, the household in the control group who has the value of propensity score nearest to that of the household in the replacement group is selected as the partner of matching. However, in this method, if the value of the propensity score of the control group household is far from that of the replacement group household, the quality of matching decreases. To prevent this issue, the radius matching sets an upper limit on the value of propensity score for matching and targets all control households whose propensity scores fall within the certain range. Kernel matching uses the kernel function and constructs the counterfactual dependent variable. As this method targets almost all control group households, the size of variance is smaller than either of the other two matching methods (in the opposite direction, kernel matching makes the largest bias).

Similar to the estimation based on regression analysis, in the estimation of causality effect based on propensity score matching, it is assumed that the assignment of treatment depends only on observable variables. Therefore, if unobservable factors affect both the assignment of treatment and the dependent variable (potential outcome), propensity score matching estimators become biased estimators. Here, in the case of our study, the decision of replacement (assignment of treatment) is not always based on only observable covariates. That is, it is possible that the replacement of an energy-consuming air conditioner with an energy-efficient one depends on not only observable covariates but also unobservable ones. In this case, using only the PSM method does not guarantee the obtainment of the estimated causality effect by complementing the counterfactual adequately. One of the techniques to handle this case is the DD method.

6.2.3. Difference-in-Differences

The difference-in-differences (hereafter, DD) method allows the analyst to estimate the causality effect by excluding the influences of unobservable covariates. For conducting the DD method, the data need to fulfil two conditions. First is the existence of treatment and control groups, each with more than one individual. Second is that we can observe the data at two time points, before and after the assignment of the treatment. Here, we assume that T is the treatment group and C is the control group, and 1 indicates the pre-treatment state and 2 the post-treatment state. In this instance, our dataset is represented in Table 6.1.

Kenichi Mizobuchi and Hisashi Tanizaki

Table 6.1. Difference-in-differences

Time point	T (Treatment group)		C (Control group)	
Time 1	(treatment) Y1T(1)	(no treatment) Y0T(1)	(treatment) Y1C(1)	(no treatment) Y0C(1)
Time 2	(treatment) Y1T(2)	(no treatment) Y0T(2)	(treatment) Y1C(2)	(no treatment) Y0C(2)

Note: The underlined part indicates an unfulfilled potential outcome, and the other part indicates a fulfilled potential outcome.

Each causality effect of treatment and control group at time 2 is defined as follows:

$$\text{Causality effect on treatment group (time 2)} = Y_{1T}(2) - Y_{0T}(2) \quad (6.5)$$

$$\text{Causality effect on control group (time 2)} = Y_{1C}(2) - Y_{0C}(2) \quad (6.6)$$

However, there are unfulfilled outcomes in either equations (Y0T(2) in Equation (6-5) and Y1C(2) in Equation (6-6)), and thus, we cannot estimate the causality effect based on either of these equations. This study employs average treatment on the treated (i.e., ATT), and thus, in Equation (6-5), if a household in the treatment group did not replace its air conditioner, we need to complement the counterfactual Y0T(2) with whatever happened at time 2. This complementary method is the DD.

For estimating the ATT based on DD, we need to follow two steps. First, we take the difference between the observable outcomes of time 1 and time 2 for both the treatment and the control groups (i.e., $Y_{1T}(2) - Y_{0T}(1)$ and $Y_{0C}(2) - Y_{0C}(1)$). Second, we find further differences between them as follows:

$$ATT_{DD} = [Y_{1T}(2) - Y_{0T}(1)] - [Y_{0C}(2) - Y_{0C}(1)] \qquad (6.7)$$

This is the ATT based on DD, and Figure 6.2 represents the ATTDD visually. Here, for complementing the counterfactual of Y0T(2) and examining the causality effect of Y1T(2)-Y0T(2), the assumption of the following parallel trends of treatment and control groups are as follows:

$$Y_{0T}(2) - Y_{0T}(1) = Y_{0C}(2) - Y_{0C}(1)$$

This parallel trend implies that if households of treatment group were not assigned the treatment (i.e., they did not replace their air conditioners), they made a shift from time 1 to time 2 similar to households of the control group. If this assumption is correct, DD can estimate the ATT by excluding the unobservable covariates.

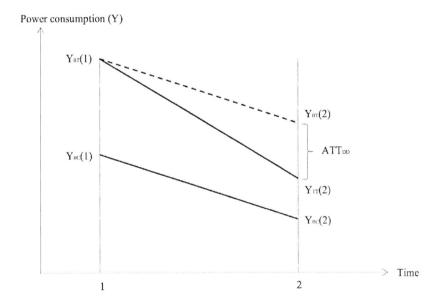

Figure 6.2. ATT of Difference-in-differences method.

However, the assumption of parallel trends is not always accurate, and thus, the estimation of ATT by the DD may be biased. In particular, the characteristics of households between the treatment and control groups may be significantly different, and thus, the parallel trend assumption may not be satisfied. Therefore, we combined DD method with the PSM method that draws households from the control group similar to those from the treatment group. Consequently, the parallel trend assumption would become easier to correct. Therefore, this study estimates the power-saving effect of replacing with an energy-efficient air conditioner by combining the DD and PSM methods.

6.3. DATA

This study targeted households residing in Kansai area, which comprises six prefectures (Osaka, Kyoto, Hyogo, Nara, Shiga, and Wakayama). Most of the households in Kansai area purchase electricity from the KEPCO (Kansai Electric Power Co., Inc.)[22]. For ease of access, this electric company provides customers with online data on their respective monthly electricity consumption for the two-year period prior to the study (2013–2014), which we requested the participant households to submit for our analysis[23]. This study commissioned a questionnaire survey to a web research company, and asked households who live in Kansai area whether they purchased a replacement or additional energy-efficient air conditioner in the previous two years, and if so, when and how many; their socio-economic characteristics; the number of other electric appliances they possess; and so on. We conducted this study in two instalments, namely the preliminary survey in February 2015 and the main survey in April 2015. From these two

---

[22] http://www.kepco.co.jp/.
[23] It is provided in the form of an Excel file. These data include not only monthly electricity consumption data but also the date of meter reading, number of days of utilization, charge for use, carbon dioxide emission, and so on.

surveys, we could obtain both monthly electricity consumption data[24] and questionnaire survey data for 733 households. We exclude those households that have installed photovoltaic systems from our dataset, because the amount of electricity consumption generated by home photovoltaic systems are not included in monthly electricity consumption data provided by KEPCO. Table 6.2 shows the descriptive statistics of our data.

Table 6.2. Descriptive statistics

Variable	Obs.	Mean	Std. Dev	Min	Max
Treatment dummy	660	0.142	0.35	0	1
# Air conditioner	660	2.758	1.701	0	9
# TV	660	2.980	1.170	1	8
# Refrigerator	660	2.162	0.467	1	5
Dishwasher 1)	660	0.342	0.636	0	2
Washing machine 2)	660	0.332	0.599	0	2
Electric Kettle 3)	660	0.500	0.734	0	2
Age	660	50.371	10.547	20	69
Income 4)	660	3.497	1.714	1	8
# Family member	660	2.650	1.340	1	9
Children (dummy)	660	0.100	0.300	0	1
Elderly (dummy)	660	0.258	0.438	0	1
Singlehood (dummy)	660	0.298	0.458	0	1
Double income (dummy)	660	0.294	0.456	0	1
Homeownership (dummy)	660	0.724	0.447	0	1
Floor size 5)	660	3.282	1.354	1	6
Detached (dummy)	660	0.498	0.500	0	1

1) 0: no dishwasher; 1: low usage frequency; 2: high usage frequency. 2) 0: no cloth washer; 1: low usage frequency; 2: high usage frequency. 3) 0: no electric kettle; 1: low usage frequency; 2: high usage frequency. 4) 1: under 200 million yen; 2: 200–399 million yen; 3: 400–599 million yen; 4: 600–799 million yen; 5: 800–999 million yen; 6: 1,000–1,199 million yen; 7:1,200–1,399 million yen; 8: over 1,400 million yen. 5) 1: under 30 m2; 2: 31–60 m2; 3: 61–90 m2; 4: 91–120 m2; 5: 121–150 m2; 6: over 151 m2.

---

[24] The time period of our electricity consumption data is from April 2013 to January 2015, which is an overlapping term of both first and second researches.

This study regards households who switched to energy-efficient air conditioners as the treatment group and those who did not do so as the control group, and it investigates the power-saving effect of replacing the air conditioner. Here, the variable of "Treatment dummy" is the replacement variable. For example, certain households take "1" in this variable, which implies that although this household had not replaced its air conditioner in 2013, it has completed the replacement in 2014. Here, with the exception of with or without the replacement, we also need to consider other factors that affect a household's electricity consumption. For example, socio-economic variables such as household income, number of people in the household, status of possession and size of the house, and possession situation and usage condition of the electric appliances. Moreover, outdoor air temperature can be a significant factor influencing the usage condition of the air conditioner. Figure 6.3 shows the average outdoor temperature of each month in 2013 and 2014. From this Figure, in summer (i.e., July, August, September), we confirm that the average outdoor temperatures of 2013 are higher than those of 2014.

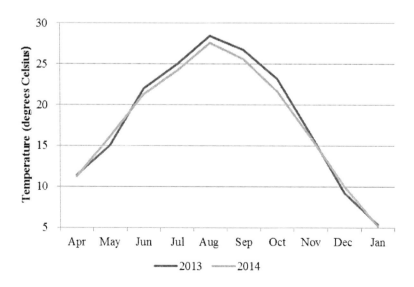

Figure 6.3. Average outdoor temperature in 2013 and 2014.

Table 6.3 shows the average monthly electricity consumption data (kWh/day)[25] for each year (2013 and 2014) and each month (from April to January). For the replacement households, 2013 is considered as the period before the replacement and 2014 is considered as the period after the replacement. Therefore, we can compare the monthly household electricity consumption between two time points. This Table shows two findings. (1) Power consumption of replacement households (treatment group) is larger than that of the control group's households for all months. This finding may imply that household characteristics are significantly different between the replacement group and the control group. (2) These differences (i.e., the size of difference of average monthly power consumption between the replacement and control groups) decrease in all months from 2013 to 2014. In particular, although we can confirm the significant differences for April, July, August, and September 2013, only that of August is statistically significant in 2014. This decrease of difference might be the replacement effect of the energy-efficient air conditioner. However, these results did not consider the effects of observable and non-observable covariates of households, and thus, we cannot confirm that this is the causality effect of switching to energy-efficient air conditioners. Therefore, we verify the causality effect by the combined method of DD and PSM as shown in Section 6.2.

First, we estimate the propensity score based on the probit estimation. We denote the replacement households with "1" and the others with "0," and then, we estimate the probit model. The explanatory variables are age, number of family members, number of children in elementary schooling, number of elderly people aged over 65 years, household income, marital status, whether double income, status of possession of the house, nature of the house (independent, apartment, and so on), number of air conditioners, number of television

---

[25] Although we can get electricity consumption data in increments of month, we also know the number of days of utilization for each month. Therefore, we calculated the daily power consumption of each month.

sets, number of refrigerators, frequency of use of dish washer, frequency of use of electric clothes dryer, and outdoor temperature in 2013. Based on the estimated propensity score, we conducted three types of matching methods, which are NN matching, radius matching, and kernel matching, as we showed in Section 6.2. Finally, we get the ATT of replacement air conditioners by using the DD method.

Table 6.3. Electricity consumption (kWh/day)

	Group	Electricity(2013)	Difference		Electricity (2014)	Difference	
April	Replacement	15.565			14.573		
	Control	12.813	2.751	*	12.637	1.936	
May	Replacement	13.071			11.447		
	Control	11.262	1.808		10.585	0.862	
June	Replacement	11.383			10.898		
	Control	10.539	0.844		10.158	0.740	
July	Replacement	12.991			11.360		
	Control	11.572	1.419		10.673	0.687	
August	Replacement	17.882			15.399		
	Control	14.202	3.681	**	13.044	2.355	**
September	Replacement	15.624			13.253		
	Control	13.055	2.569	**	11.751	1.502	
October	Replacement	12.430			11.391		
	Control	10.678	1.752	*	10.037	1.353	
November	Replacement	12.493			11.962		
	Control	10.981	1.511		10.518	1.444	
December	Replacement	15.829			15.507		
	Control	13.405	2.424		12.939	2.567	
January	Replacement	16.587			17.633		
	Control	16.075	0.512		16.159	1.474	

***, **, and * indicate statistical significance at 1%, 5%, and 10% respectively.

Table 6.4. Means of covariates between the replacement and control group before and after Kernel matching

Covariates		Unmatched	Mean			
		Matched	Replacement	Control	Difference	t-stat.
air_n	the number of air-conditioner	U	3.578	2.669	0.909	4.800 ***
		M	3.578	3.440	0.138	0.300
age	age	U	54.233	50.006	4.227	3.570 ***
		M	54.233	53.681	0.552	0.330
people	number of family member	U	2.922	2.624	0.298	1.970 **
		M	2.922	2.943	-0.021	-0.120
child	number of children	U	0.089	0.099	-0.010	-0.300
		M	0.089	0.109	-0.020	-0.410
old	number of old people	U	0.333	0.249	0.085	1.700 *
		M	0.333	0.333	0.000	0.000
single	0/1 dummy	U	0.167	0.314	-0.147	-2.860 ***
		M	0.167	0.178	-0.012	-0.170
double_ income	0/1 dummy	U	0.233	0.314	-0.081	-1.340
		M	0.233	0.303	-0.069	-0.010
income	income	U	3.711	3.460	0.251	1.290
		M	3.711	3.773	-0.062	-0.210
own	0/1 dummy	U	0.944	0.738	0.207	4.930 ***
		M	0.944	0.929	0.015	0.390
wide	wide of house	U	3.722	3.230	0.492	3.240 ***
		M	3.722	3.655	0.067	0.230
detached	0/1 dummy	U	0.656	0.480	0.175	3.090 ***
		M	0.656	0.621	0.034	0.420
tv_n	number of tv	U	3.556	2.910	0.645	4.910 ***
		M	3.556	3.466	0.090	0.330
ref_n	number of refrigerator	U	2.211	2.151	0.060	1.150
		M	2.211	2.188	0.023	0.210
dish_w	frequency of using dish washer	U	0.389	0.331	0.058	0.800

Table 6.4. (Continued)

Covariates		Unmatched Matched	Mean Replacement	Control	Difference	t-stat.
cloth_w	frequency of using cloth washer	M	0.389	0.386	0.003	-0.040
		U	0.433	0.312	0.121	1.790 *
pot	frequency of using pot	M	0.433	0.415	0.019	0.010
		U	0.567	0.497	0.069	0.830
temp_13	monthly outdoor air temperature of August, 2013	M	0.567	0.554	0.013	0.170
		U	18.090	18.305	-0.215	-1.600
		M	18.090	18.217	-0.127	-0.470

***, **, * indicates statistically significance at 1%, 5%, and 10% respectively.

6.4. ESTIMATION RESULTS

6.4.1. Propensity Score Estimator and the Balance of Matching

We used STATA ver.11 for matching estimation. Table 6.4 shows the estimation results of the probit model, and we only show that of using kernel matching. From this table, the number of air conditioners, age, number of family members, possessed status of the house, size of the house, detached house, and the number of television sets have a negatively significant effect for switching to energy-efficient air-conditioner before matching. On the other hand, we can confirm that any covariates do not have significant effect for the replacement after matching.

Table 6.5 shows the extent of decrease in bias before and after implementation of the matching of covariates between replacement group and control group households. From this table, although the biases of all three matchings decrease, the diminution of biases of

Radius and Kernel matching is larger than that of NN matching. Considering the result of Kernel matching, although we can confirm that the size of mean bias before matching is 27.6% (median bias is 19.5%), it decreased to 3.7% (3.7%) after matching and was not significant. Therefore, from the results of Tables 6.4 and 6.5, we can say that the differences of covariates between the replacement and the control group were adjusted suitably.

Table 6.5. Balance check of covariates before and after matching

	mean bias (%)	median bias(%)	LR chi2	p > chi2
NN				
unmatched	27.6	19.5	59.34	0.000
matched	9.0	9.1	11.3	0.8
Radius				
unmatched	27.6	19.5	59.34	0.000
matched	3.3	3.1	0.81	1.000
Kernel				
unmatched	27.6	19.5	59.34	0.000
matched	3.7	3.7	1	1.000

6.4.2. Estimation Results of ATT with Combined DD and Matching

Tables 6.6 shows the estimation results of ATT based on Equation (6.7), which is the DD estimator after propensity score matching, and also shows the result of each matching method (NN matching, Radius matching, and Kernel matching). The left-hand side of the table shows the results before matching and the right-hand side shows those after matching. Each of these sides consists of three columns. The first column represents the difference in average household electricity consumption (kWh/day) from 2013 to 2014 of each replacement and control group. The second column represents the difference between tow values of the first column (i.e., ATT or DD estimator after matching), and the third column represents t-statistic of the ATT estimator of the second column.

Table 6.6. Electricity-saving effects of replacement (DD+PSM)

	Group	difference of electricity (U)	DD	t-stat.	difference of electricity (M)	DD	t-stat.
NN	Replacement	-1.299			-1.299		
	Control	-0.692	-0.607	-3.170 ***	-0.938	-0.361	-1.140
Radius	Replacement	-1.299			-1.299		
	Control	-0.692	-0.607	-3.170 ***	-0.946	-0.353	-1.470
Kernel	Replacement	-1.299			-1.299		
	Control	-0.692	-0.607	-3.170 ***	-0.954	-0.345	-1.440

***, **, * indicates statistically significance at 1%, 5%, and 10% respectively.

From these tables, it can be observed that the DD estimators before matching become negative and statistically significant. This result is consistent with the expected result from Table 6.3, where the household who replaced its air conditioner with an energy-efficient one can decrease power consumption compared with the control group's household. However, as we confirmed from Table 6.4, some observable covariates between replacement and control group's households had significant difference before matching, and these covariates might affect the electricity consumption of households. Therefore, to exclude the differences of these observable covariates, we need to conduct the matching and analyze its results. The DD estimators after matching are shown on the right-hand side of Tables 6.6. Based on the 5% significant level, DD estimators are negative but not significant for all matching methods. Moreover, in almost all months, the DD estimators after matching decrease more than those before matching. This means that the observable covariates have an influence, and in such cases, the assumption of parallel trends, which we showed in Figure 6.2, is not satisfied. From these estimation results of ATT based on the combined DD and PSM methods, we can confirm that the replacement of air conditioners to energy-efficient ones did not contribute to the decrease of household power consumption.

6.4.3. The Effect of Observable and Unobservable Covariates

As we mentioned in Section 6.2, the matching estimator based on the propensity score assumes that the assignment of replacement is dependent only on the observable covariates. Therefore, if unobservable covariates affect the assignment of the replacement, the matching estimator becomes a biased estimator. To exclude the influences of unobservable factors and decrease the size of bias, this study attempted to estimate ATT by combining the PSM and DD methods. Table 6.7 compares the ATT estimator of DD with PSM and the ATT estimator of only PSM. That is, Table 6.7 shows each ATT estimator of Table 6.6 in the first column, and the ATT of only matching on the extreme right (i.e., Y1T(2) – Y0C(2) of Figure 6.2, where Y0C(2) is closer to Y0T(2) after matching). For each matching method (NN matching, Radius matching, Kernel matching), the ATT estimators based on DD with matching take similar values. On the other hand, in the cases of April and September, although the size of ATT is similar in Radius and Kernel matching, the size of ATT in NN matching is larger than them.

Table 6.7. ATT of replacement (kWh/day)

	no matching		NN	Radius	Kernel	Difference between replacement and control after matching in 2014
	DD		DD+PSM			PSM
Electricity saving (kWh/day)	-0.607	***	-0.361	-0.353	-0.345	-2.650

***, **, and * indicate statistical significance at 1%, 5%, and 10% level, respectively.

In addition, we can confirm that the ATT estimator with only PSM is not significant. Moreover, the size of ATT based only on PSM differs from that based on DD with PSM. This result can be considered as the bias from the influence of unobservable factors.

CONCLUSION

This study examined the causality effect from the shift to energy-efficient air conditioners to decrease power consumption, by comparing the replacement and non-replacement household groups. Based on the questionnaire survey and electricity consumption data for previous two years (2013–2014) of 733 participant households who live in Kansai area, we combined the DD and PSM methods, and estimated the ATT of the replacement. We assumed the "replacement households" who replaced their air conditioners with energy-efficient ones for the previous two years as the treatment group, and the "non-replacement households" as the control group. From our empirical analyses, we could not confirm the significant ATT effect of the replacement.

In general, we can expect that if households only replace their electric appliances with new energy-efficient ones without changing their consumption behavior, they can save power consumption by replacement. Based on this expectation, policymakers must encourage the replacement of household electric appliances (e.g., air conditioner, refrigerator, and washing machine) with more energy-efficient ones through means such as a subsidy policy. However, some scholars assert that such replacement may induce additional energy consumption, which is known as the rebound effect[26]. In our study, no significant

---

[26] Although technological progress can reduce energy usage, reduction in the cost of energy services may lead to an increase in the demand for energy services, and this additional energy demand may completely (or partially) offset the expected energy saving from the technological progress. This phenomenon is called the rebound effect (Sorrell and Dimitropoulos 2008).

ATT effect may have been brought by this rebound effect. Improvement of household thermal insulation performance is one of the effective countermeasures to decrease the rebound effect. However, the Japanese government had not sufficiently focused on improving household insulation previously. Therefore, if we consider the present situation that the improvement of energy efficiency performance of electric appliances has hit a peak in Japan, encouraging the increase of well-insulated houses may be the next effective measure for energy efficiency of the household sector.

Moreover, we showed the possibility that if we estimate ATT with only the DD method, we overestimate the ATT estimator (i.e., the assumption of parallel trend does not hold because of the different covariates between the treatment and control groups). Therefore, when we examine the causality effect from the replacement with energy-efficient equipment, there is a need to estimate it by combining the PSM and DD methods.

Chapter 7

SUMMARY AND CONCLUSION: DIRECTIONS FOR FUTURE POWER-SAVING POLICIES

ABSTRACT

This chapter briefly summarizes the results verified in each chapter of this book. Based on the results and considering the current energy situation in Japan, we propose various measures to promote energy-saving behavior among Japanese households in the future from the perspectives of cost effectiveness, policy effectiveness, economic welfare, and so on.

Keywords: power saving, environmental policy, CO_2 emissions

7.1. SUMMARY OF THIS BOOK

In Japan, although energy conservation measures aimed at reducing the effects of global warming advanced, the supply of electricity was drastically reduced due to the shutdown of nuclear power plants after the Great East Japan Earthquake in March 2011. Therefore, further measures for energy saving are needed. In particular, in the household sector, where regulation is difficult, government measures are not adequate. In each chapter of this book, valid measures to encourage power saving and their effects were examined for Japan's household sector using econometric analysis and field experiments.

In Chapter 1, we reviewed the time series trends in CO_2 emissions and energy consumption in the Japanese household sector, and found that they have been rapidly increasing over the last 25 years. Furthermore, among the energy sources, it was confirmed that the proportion of electric power in the final energy consumption is increasing due to the diversity of household appliances and increase in ownership rate. In addition, we summarized the preceding research on measures to promote energy conservation behavior in the household sector.

In Chapter 2, we estimated the AIDS model, which is a household demand system model, and examined the demand behavior of Japanese households. Both expenditure elasticity and price elasticity of energy demand were significant, taking positive and negative values, respectively. In particular, the price elasticity (in absolute value) was greater than one; thus, the household sector was expected to respond highly to economic incentives (subsidies, penalty charges, and so on). We confirmed this in field experiments explained in Chapters 4 and 5. Furthermore, after the Great East Japan Earthquake, structural changes were confirmed in the household demand system, and the magnitude of expenditure elasticity changed.

In Chapter 3, the effect of the government's request for power saving to the household sector was examined by means of quantitative analysis using the monthly electricity expenditure data (January 2003 to December 2016). The results suggested that after the Great East Japan Earthquake of March 2011, a significant power-saving effect was confirmed in the area where the government request for power saving was issued. Furthermore, as the consciousness of energy conservation increased nationwide, significant power-saving effects were observed even in some areas where no request was made. Even after 2013, when power-saving requests ceased to be issued, significant power-saving effects were observed in some areas, indicating that the power-saving behavior is getting established in the household sector.

In Chapter 4, based on the results of Chapter 2, we examined whether economic incentives encourage power-saving behavior, using field experiments. For 54 households in Matsuyama-shi, Ehime Prefecture, Japan, we set up stepwise compensation according to the energy-saving rate for the three months from November 2010 to January 2011. About 34% of the participating households succeeded in saving electricity. On the other hand, as a result of the questionnaire survey, it was speculated that the marginal cost of energy-saving behavior was high, as the number of people who responded negatively with regard to energy-saving behavior doubled even if the amount of remuneration was doubled.

In Chapter 5, field experiments were used to examine the impact of economic factors (compensation) and psychological factors (comparison with others) on households' energy-saving behavior. We surveyed 236 households in Matsuyama City, Ehime Prefecture, which were divided into three groups: (i) Reward group; ii) Reward and Comparative Feedback group; and iii) Control group. We compared the power-saving rates in the two months between October and November 2011. The results clarified that both economic and psychological factors promote power-saving behavior. Furthermore, people tend to demonstrate energy-saving behavior when others around do so. Furthermore, from the results of the preliminary questionnaire and post-questionnaire, it was clarified that households underestimate the marginal cost of energy-saving behavior.

In Chapter 6, we examined whether the adoption of energy-efficient household appliances will help save electricity. Specifically, we compared the electricity usage of households that have replaced their energy-consuming air conditioners with energy-efficient ones within the past two years and those that have not done so. The target respondents comprised 733 households living in the Kansai area, who were asked to submit monthly electricity consumption data for the past two years, and answer the questionnaire regarding household attributes, among others. We analyzed the results by using a combination of propensity score matching and difference-in-differences methods. Our results suggested that switching to energy-efficient air conditioners did not bring about a power-saving effect.

7.2. DIRECTIONS FOR FUTURE POWER-SAVING POLICIES

This book has examined the ways to encourage power-saving behavior among households. Each chapter clarified that there are several effective means to promote energy saving (government request, economic incentive, psychological means, and so on). As of December 2017, several nuclear power plants have recommenced operations, and the problem of power supply shortage is being resolved. On the other hand, however, the Government of Japan has set 26% as the target for reducing greenhouse gas emissions by 2030 compared to 2013, of which the goal is to reduce the energy consumption of the household sector by 39.3% compared to 2013.

The two issues related to energy conservation policies targeting households are as follows: 1) difficulties in imposing penalties and regulations; and 2) cost effectiveness. First, as compared with other sectors, such as the industrial sector and the transport sector, the household sector has had few experiences of energy reduction measures using penal regulations or regulations so far. Thus, it is anticipated that there will be major opposition to the introduction of such a policy in this sector. However, as we clarified in Chapter 3, the temporary power-saving request by the government was effective, and the spread of power-saving behavior and continuity were observed. An important point to consider during the implementation of this policy would be to explain the policy objectives firmly to people. It is considered that the power-saving request implemented against the power supply shortage brought about by the shutdown of nuclear power plants owing to safety was widely accepted by people, which is the reason for the effect of request for power saving. Meanwhile, it is difficult to introduce penalties; however, as shown in the field experiments in Chapters 4 and 5, there is a possibility that energy conservation behavior can be promoted among the households by way of compensation. However, the difficulty of introducing such policies is that it is necessary to clarify from what level to save electricity. If based on the previous year, it is disadvantageous for households who have already demonstrated power-saving behavior before that, while it will be advantageous for households who did not do so. Thus, we need to consider fair standards for electricity saving.

Second, regarding cost effectiveness, it is difficult to obtain expensive means in a country like Japan, where the debt of the government is prominent among developed countries. Therefore, it is difficult to introduce the compensation system as shown in Chapters 4 and 5, rather than thinking that it is realistic to use the government request and psychological means, as shown in Chapters 3 and 5. Furthermore, as clarified in Chapters 4 and 5, households underestimated the marginal cost of energy-saving behavior. Regarding this point, providing information on highly efficient energy-saving measures is expected to be a cost-effective policy with a large energy-saving effect expected. In addition, a subsidy system for disseminating energy-saving products attracts attention (Matsumoto 2015). Since a subsidy policy has multiple goals, such as economic revitalization and acceleration of the spread of new technologies as well as a single goal of energy consumption reduction, the possibility of policy implementation is also high.

The enjoyment of energy services using electricity increases the utility of households, but causes an increase in CO_2 emissions and a problem of energy shortage. For this reason, the government is required to have means to resolve the environmental problems without impairing the welfare of households. As verified in this book, we showed some effective means to save electricity, yet we could not fully verify the sustainability of the effect and cost effectiveness. Therefore, it is necessary to verify these measures in consideration of policy effectiveness in practice.

ACKNOWLEDGMENTS

This research is supported by the Japan Society for the Promotion of Science (Grant-in-Aid for Scientific Research (C) #17K03742 and #17K03657).

REFERENCES

Abrahamse, W., Steg, L., Vlek, C., Rothengatter, T., 2005. A review of intervention studies aimed at household energy conservation. *Journal of Environmental Psychology*, 25, 273-291.

Allcott, Hunt. 2011. "Social Norms and Energy Conservation." *Journal of Public Economics* 95:1082–1095.

Alston Julian M, Foster Kenneth A, Green Richard D. 1994. "Estimating Elasticities with the Linear Approximate Almost Ideal Demand System: Some Monte Carlo Results." *Review of Economics and Statistics* 76 (22):351-356.

Arikawa, Hiroshi., Cao, Yang., and Matsumoto, Sigeru. 2014. "Attitudes about Nuclear Power and Energy-saving Behavior Among Japanese Households." *Energy Research and Social Science* 2: 12–20.

Becker, Lawrence J. 1978. "Joint Effect of Feedback and Goal Setting on Performance: A Field Study of Residential Energy Conservation." *Journal of Applied Psychology* 63:428–433.

Brandon, Gwendolyn, and Alan Lewis. 1999. "Reducing Household Energy Consumption: A Qualitative and Quantitative Field Study." *Journal of Environmental Psychology* 19:75–85.

Buse, Adolf. 1994. "Evaluating the Linearized Almost Ideal Demand System." *American Journal of Agricultural Economics* 76(4):781–793.

Buse, Adolf. 1998. "Testing Homogeneity in the Linearized Almost Ideal Demand System." *American Journal of Agricultural Economics* 80(1):208–220.

Buse, Adolf, and Wing Chan H. 2000. "Invariance, Price Indices and Estimation in Almost Ideal Demand Systems." *Empirical Economics* 25(3):519–539.

Clark, Christopher F., Matthew J. Kotchen, and Michael R. Moore. 2003. "Internal and External Influences on Pro-environmental Behavior: Participation in a Green Electricity Program." *Journal of Environmental Psychology* 23:237–246.

Deaton, Angus, and John Muellbauer. 1980. "An Almost Ideal Demand System." *American Economic Review* 70:312–326.

Davis, Lucas W. 2008. "Durable Goods and Residential Demand for Energy and Water: Evidence from a Field Trial." *RAND Journal of Economics* 39: 530–546.

Davis, Lucas W., Alan Fuchs, and Paul Gertler. 2014. "Cash for Coolers: Evaluating a Large-Scale Appliance Replacement Program in Mexico," *American Economic Journal: Economic Policy* 6(4): 207–238.

Deci, Edward L. 1971. "Effects of Externally Mediated Rewards on Intrinsic Motivation." *Journal of Personality and Social Psychology* 18:105–115.

Deci, Edward L., Richard Koestner, and Richard M. Ryan. 1999. "A Meta-analytic Review of Experiments Examining the Effects of Extrinsic Rewards on Intrinsic Motivation." *Psychological Bulletin* 125:627–668.

Dubin, Jeffrey A., Allen K. Miedema, and Ram V. Chandran. 1986. "Price Effects of Energy-efficient Technologies: A Study of Residential Demand for Heating and Cooling." *RAND Journal of Economics* 17: 310–325.

EDMC. 2015. *Handbook of Energy & Economic Statistics*. The Institute of Energy Economics, Quantitative Analysis Unit, Tokyo, Japan (in Japanese).

Efron, Bradley, and Tibshirani, Robert, J. 1993. *An Introduction to the Bootstrap. Monographs on Statistics and Applied Probability*. CHAPMAN&HALL/CRC.

Ek, Kristina, and Patrik Soderholm. 2008. "Norms and Economic Motivation in the Swedish Green Electricity Market." *Ecological Economics* 68:169–182.

Ek, K., Soderholm, P., 2010. The devil is in the details: Household electricity saving behavior and the role of information. *Energy Policy*, 38, 1578-1587.

Electricity Supply and Demand Verification Subcommittee. 2013. "*Electricity Supply and Demand Verification Subcommittee Report*." (in Japanese) http://www.meti.go.jp/committee/ sougouenergy/sougou/jukyu_kensho/pdf/report01_02_00.pdf.

Feenstra, Robert C., and Marshall B. Reinsdorf. 2000. "An Exact Price Index for the Almost Ideal Demand System." *Economics Letters* 66:159–162.

Filippini, Massimo. 1995. "Electricity Demand by Time of Use: An Application of the Household AIDS Model." *Energy Economics* 17(3):197–204.

Fitzenberger, Bernd. 1997. "The Moving Blocks Bootstrap and Robust Inference for Linear Least Squares and Quantile Regressions." *Journal of Econometrics* 82:235–287.

Freedman, David A., and Stephen Peters C. 1984. "Bootstrapping a Regression Equation: Some Empirical Results." *Journal of American Statistical Association* 79(385):97–106.

Frey, Bruno S., and Reto Jegen. 2001. "Motivation Crowding Theory." *Journal of Economic Surveys* 15:589–611.

Fujimi, Toshio, and Stephanie E. Chang. 2014. "Adaptation to Electricity Crisis: Businesses in the 2011 Great East Japan Triple Disaster." *Energy Policy* 68: 447–457.

Goldman, Charles A., Joseph H. Eto, and Galen L. Barbose. 2002. *"California Customer Load Reductions during the Electricity Crisis – Did they Help Keep the Lights On?"* Ernst Orlando Lawrence Berkeley National Laboratory LBNL-49733, http://emp.lbl.gov/sites/all/files/lbnl%20-%2049733.pdf.

Green, Richard, David Rocke, and William Hahn. 1987. "Standard Errors for Elasticities: A Comparison of Bootstrap and Asymptotic Standard Errors." *Journal of Business and Economic Statistics* 5(1):145–149.

Green, Richard, and Julian Alston M. 1991. "Elasticities in AIDS Models: A Clarification and Extension." *American Journal of Agricultural Economics* 73(4):874–875.

Greene, William H. 2008. *Econometric Analysis*. New Jersey: Prentice Hall.

Hahn, William F. 1994. "Elasticities in AIDS Models: Comments." *American Journal of Agricultural Economics* 76:972–977.

Hashimoto, Noriko, 2004. *Changing Japanese Society and Behavior of Household Consumption: Analysis with the AIDS Model*. Osaka: Kansai University Publishing (in Japanese).

Jeroen, C. J. M., and van den Bergh. 2008. "Environmental Regulation of Households: An Empirical Review of Economic and Psychological Factors." *Ecological Economics* 66:559–574.

Joskow, Paul L., and Donald B. Marron. 1992. "What Does a Negawatt Really Cost? Evidence from Utility Conservation Programs." *Energy Journal* 13: 41–74.

Kansai Electric Power Company. 2014. *"About the Power Situation this Summer."* (in Japanese) http://www.kepco.co.jp/corporate/pr/2014/__icsFiles/afieldfile/2014/08/28/0828_1j.pdf.

Kotchen, Matthew J., and Michael R. Moore. 2007. "Private Provision of Environmental Public Goods: Household Participation in Green-electricity Programs." *Journal of Environmental Economics and Management* 53:1–16.

Krinsky, Itzhak, and A. L. Robb. 1991. "Three Methods for Calculating the Statistical Properties of Elasticities: A Comparison." *Empirical Economics* 16(2):199–209.

Künsch, Hans R. 1989. "The Jackknife and the Bootstrap for General Stationary Observations." *Annals of Statistics* 17: 1217–1241.

Leighty, Wayne, and Alan Meier. 2011 "Accelerated Electricity Conservation in Juneau, Alaska: A Study of Household Activities that Reduced Demand 25%." *Energy Policy* 39(5): 2299–2309.

Liu, Regina Y., and Singh, Kesar. 1992. "Moving Blocks Jackknife and Bootstrap Capture Weak Dependence." In: *Exploring the Limits of the Bootstrap*, edited by Lepage R and Billiard L, 224–248. New York: Wiley.

MacKinnon, James G. 2006. "Bootstrap Methods in Econometrics." *Economic Record* 82:S2–S18.

Matsumoto, Shigeru. 2015. *Environmental Subsidies to Consumers – How Did They Work in the Japanese Market?* Rondon and New York: Routledge.

McClelland, Lou, and Stuart W. Cook. 1980. "Promoting Energy Conservation in Master-metered Apartment through Group Financial Incentives." *Journal of Applied Social Psychology* 10 (1):20–31.

Midden, Cees J. H., Joanne F. Meter, Mieneke H. Weenig, and Henk J. A. Zieverink. 1983. "Using Feedback, Reinforcement and Information to Reduce Energy Consumption in Households: A Field-experiment." *Journal of Economic Psychology* 3 (1):65–86.

Ministry of Economy, Trade and Industry. 2013. "*Summer Electricity Saving Menu.*" (in Japanese) http://www.kansai.meti.go.jp/3-9kaihatsu/electricity/2013_taisaku_katei.pdf.

Mizobuchi, Kenichi. 2008. "An Empirical Study on the Rebound Efect Considering Capital Costs." *Energy Economics* 30:2486-2516.

Mizobuchi, Kenichi, and Takeuchi, Kenji. 2012. "Using Economic Incentive to Conserve Electricity Consumption: A Field Experiment in Matsuyama, Japan." *International Journal of Energy Economics and Policy* 2(4):318-332.

Mizobuchi, Kenichi., and Takeuchi, Kenji. 2013. "The Influences of Financial and Non-financial Factors on Energy-Saving Behavior: A Field Experiment in Japan." *Energy Policy* 63:775-787.

Moschini, Giancarlo. 1995. "Units of Measurement and the Stone Index in Demand System Estimation." *American Journal of Agricultural Economics* 77:63–68.

Moschini, Giancarlo. 1998. "The Semiflexible Almost Ideal Demand System." *European Economic Review* 42:349–364.

Natural Resources and Energy Agency. 2011. *"On Measures to Control Electricity Demand this Summer."* (in Japanese) http://www.meti.go.jp/committee/summary/0002015/014_s02_00.pdf.

Nishio, Kenichiro, and Ofuji, Kenta. 2012. *"Actual State of Energy Saving in the Summer of 2011 at Home."* Electric Power Research Institute, (in Japanese) http://criepi.denken.or.jp/jp/kenkikaku/report/detail/Y11014.html.

Nishio, Kenichiro, and Ofuji, Kenta. 2013. *"Actual State of Energy Saving in the Summer of 2012 at Home."* Electric Power Research Institute (in Japanese) http://criepi.denken.or.jp/jp/kenkikaku/report/detail/Y12026.html.

Nishio, K., and Ofuji, K. 2013. *"Ex-post Analysis of Electricity-saving Measures in the Residential Sector in the Summer of 2013"* (in Japanese), Tokyo: Central Research Institute of Electric Power Industry.

Nishio, Kenichiro, and Ofuji, Kenta. 2014. *"Actual State of Energy Saving in the Summer of 2013 at Home,"* Electric Power Research Institute (in Japanese) http://criepi.denken.or.jp/jp/kenkikaku/report/detail/Y13010.html.

Nolan, Jessica M., P. Wesley Schultz, Robert B. Cialdini, B.R, Noah J. Goldstein, and Vladas Griskevicius. 2008. "Normative Social Influence is Underdetected." *Personality and Social Psychology Bulletin* 34:913–923.

Nyborg, Karine, Richard B. Howarth, and Kjell A. Brekke. 2006. "Green Consumers and Public Policy: On Socially Contingent Moral Motivation." *Resource and Energy Economics* 28:351–366.

Ofuji, Kenta, and Nishio, Kenichiro. 2011. "California Supply and Demand Warning FlexAlert and its Evaluation of Power Saving Effect." Electric Power Research Institute, *SERC Discussion Paper* (in Japanese) 11019, http://criepi.denken.or.jp/jp/serc/discussion/download/11019dp.pdf.

Ofuji, K, and Kimura, O. 2011. "About the 20/20 Electricity Saving Program in California and the Method of Ex-post Evaluation." Electric Power Research Institute, *SERC Discussion Paper* (in Japanese) 11011, http://criepi.denken.or.jp/jp/serc/discussion/download/11011dp.pdf.

Okajima, Shigeharu, and Hiroko Okajima. 2013. "Estimation of Japanese Price Elasticities of Residential Electricity Demand, 1990–2007." *Energy Economics* 40: 433–440.

Oladosu, Gbadebo. 2003. "An Almost Ideal Demand System Model of Household Vehicle Fuel Expenditure Allocation in the United States." *Energy Journal* 24(1):1–21.

Pashardes, Panos. 1993. "Bias in Estimating the Almost Ideal Demand System with the Stone Index Approximation." *Economic Journal* 103(419):908–915.

Petersen, John E., Vladislav Shunturov, Kathryn Janda, Gavin Platt, and Kate Weinberger. 2007. "Dormitory Residents Reduce Electricity Consumption when Exposed to Real-time Visual Feedback and Incentives." *International Journal of Sustainability in Higher Education* 8:16–33.

Population Statistics. 2009. *National Institute of Population and Social Security Research*. Cited March 29, 2009. http://www.ipss.go.jp/index-e.asp.

Reiss, Peter C., and Matthew W. White. 2008. "What Changes Energy Consumption? Prices and Public Pressures," *RAND Journal of Economics* 39(3): 636–663.

Rosenbaum, Paul R., and Donald D. Rubin. 1983. "The Central Role of the Propensity Score in Observational Studies for Causal Effects." *Biometrika* 70: 41–55.

Rossi, Nicola. 1988. "Budget Share Demographic Translation and the Aggregate Almost Ideal Demand System." *European Economic Review* 32(6):1301–1318.

Roy, Andrew D. 1951. "Some Thoughts on the Distribution of Earnings." *Oxford Economic Papers* 3: 135–145.

Rubin, Donald B. 1974. "Estimating Causal Effects to Treatments in Randomized and Nonrandomized Studies." *Journal of Educational Psychology* 66: 688–701.

Smith, Vincent H., Michael R. Kehoe, and Mary E. Cremer. 1995. "The Private Provision of Public Goods: Altruism and Voluntary Giving." *Journal of Public Economics* 58:107–126.

Sorrell, Steve, and John Dimitropoulos. 2008. "The Rebound Effect: Microeconomic Definitions, Limitations and Extensions." *Ecological Economics* 65: 636–649.

Supply and Demand Verification Committee. 2012. *"About Follow-up of Electricity Supply and Demand Countermeasures this Summer."* (in Japanese) http://www.cas.go.jp/jp/seisaku/npu/policy09/pdf/20121012/shiryo3-1-1.pdf.

Tanaka, Makoto, and Takanori Ida. 2013. "Voluntary Electricity Conservation of Households after the Great East Japan Earthquake: A Stated Preference Analysis." *Energy Economics* 39: 296–304.

Tanizaki, H., and Mizobuchi, K. 2014. "On Estimation of the AIDS Model using Moving Blocks Bootstrap and Pairs Bootstrap Methods." *Empirical Economics* 47(4):1221-1250.

Tiffin, Richard, and Magda Aguiar. 1995. "Bayesian Estimation of an Almost Ideal Demand System for Fresh Fruit in Portugal." *European Review of Agricultural Economics* 22:469–480.

Tiffin, Richard, and Kelvin Balcombe. 2005. "Testing Symmetry and Homogeneity in the AIDS with Cointegrated Data using Fully-modified Estimation and the Bootstrap." *Journal of Agricultural Economics* 56(2):253–270.

Winett, Richard A., John H. Kagel, Raymond C. Battalio, and Robin C. Winkler. 1978. "Effects of Monetary Rebates, Feedback, and Information on Residential Electricity Conservation." *Journal of Applied Psychology* 63:73–80.

Xiao, Ni, Jay Zarnikau, and Paul Damien. 2007. "Testing Functional Forms in Energy Modeling: An Application of the Bayesian Approach to U.S. Electricity Demand." *Energy Economics* 29(2):158–166.

Xie, Yu. 2013. "Population Heterogeneity and Causal Inference." *Proceedings of the National Academy of Sciences of the United States of America*, 110(16): 6262–6268.

ABOUT THE AUTHORS

Kenichi Mizobuchi
Professor
Department of Economics, Matsuyama University
Email: kmizobuc@g.matsuyama-u.ac.jp

Kenichi Mizobuchi is Professor at the Department of Economics, Matsuyama University. He earned his PhD in Economics at Kobe University. His research interest lies in environmental economics and applied econometrics, with particular focus on household energy saving behavior analysis. Based on the research results obtained, he is aiming at planning and making recommendations on highly efficient energy conservation policy for households. His empirical studies on the rebound effect by improving energy efficiency of households and research papers on field experiments on energy conservation behavior of households are published in Energy Economics, Energy Policy, and others.

Hisashi Tanizaki
Professor
Graduate School of Economics, Osaka University
Email: tanizaki@econ.osaka-u.ac.jp

Hisashi Tanizaki is a Professor at the Graduate School of Economics, Osaka University, Japan. He received the M.A. degree (1987) from Kobe University, Japan, and the Ph.D. degree (1991) from the University of Pennsylvania, USA. Since 2011, he has been a Professor at Osaka University. He is interested in econometrics and statistics, especially econometric theory and financial econometrics. He has published three books, six book chapters, and over thirty papers.

INDEX